MARS

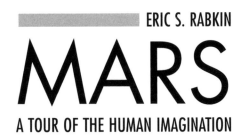

ERIC S. RABKIN

MARS

A TOUR OF THE HUMAN IMAGINATION

PRAEGER

Westport, Connecticut
London

Library of Congress Cataloging-in-Publication Data

Rabkin, Eric S.
   Mars : a tour of the human imagination / Eric S. Rabkin.
     p. cm.
   Includes bibliographical references and index.
   ISBN 0–275–98719–1 (alk. paper)
   1. Mars (Planet)—Exploration.   2. Science fiction.   3. Science in popular culture.   I. Title.
QB641.R29   2005
523.43—dc22   2005003476

British Library Cataloguing in Publication Data is available.

Library of Congress Catalog Card Number: 2005003476
ISBN: 0–275–98719–1

First published in 2005

Praeger Publishers, 88 Post Road West, Westport, CT 06881
An imprint of Greenwood Publishing Group, Inc.
www.praeger.com

Printed in the United States of America

The paper used in this book complies with the
Permanent Paper Standard issued by the National
Information Standards Organization (Z39.48–1984).

10 9 8 7 6 5 4 3 2 1

This book is for Betty,
circling the Sun with me for all our lives,
all that I could imagine
and more

CONTENTS

Color section follows page 114.

PREFACE

What is Mars? From the ancients to the present, we have imagined Mars repeatedly and studied it longingly. As our scientific knowledge of Mars has changed, our cultural imagination of Mars also has changed. The earth-centered beginnings of astronomy connected the bloody planet with the God of War. The Copernican Revolution and a later, simple mistranslation from Italian supported fantastic visions of distant Mars as the abode of life variously bizarre, ideal, or malignant. In the work of H. G. Wells and Orson Welles, Mars reflected not only eternal hopes and fears but then-current political realities. In recent years, NASAfication has brought Mars home again, imagining the Red Planet almost as an eighth continent of Earth, a candidate for exploration and exploitation both in fiction and in fact. What is Mars really, what has it been, and what may it become?

ACKNOWLEDGMENTS

Mars has been an obvious object of fascination for humankind since we first noticed its lurid glow in the ancient night. It has been an object of my fascination ever since, as a child in the 1950s, it symbolized the Red Menace both across the ocean and in the next town—or apartment! In those days, the science fictions of Mars defined our politics and our popular culture. My earnest and loving father, Joseph, habitually came home from wearying days at work to escape into the land of pulpy magazines. I thank him always for leading me to that riveting world.

I thank my students and colleagues at the University of Michigan for joining me over decades in exploring the meanings and uses of great symbols and I thank Matthew Linke of the University of Michigan Exhibits Museum for offering me the opportunity to shape a public presentation about the cultural evolution of Mars.

I thank all my fellow critics who over the years have joined me in Mars or at least in mapping it, and particularly among them Marleen Barr and George Slusser, both brilliant, both out of this world.

I join all other grateful authors in thanking those who most immediately made possible the publication of the work in hand: my friend and agent,

Donald Gastwirth; my champion and editor at Praeger, Heather Staines; and Praeger's indefatigable and tactful Alexander Andrusyszyn who made it possible to include a rich assemblage of materials that mark and are marked by Mars.

I thank also, and always, my wonderful family: David who has always played along with my enthusiasms and Rachel who cares as much about writing as anyone I know, both of whom read and improved the manuscript; Judi who is so buoyantly supportive; and my life mate, Betty, editor-in-chief and the Venus of my sky. I love you all.

Red Light in the Black Sky

From before the ancestors of humanity lifted their eyes to the sky, the red light looked down on the Earth watching, waiting, unique. On a clear night, far from the lights of human habitation, you can find that light. Gaze at it and, if you have very good eyesight and the heating and cooling of the Earth and atmosphere does not cause too much wavering, this is what you would see: a faint, pale red, impossibly distant pinprick that grows somewhat blurry as you stare. What is it?

If you do this night after night, you will see it move, move even among the other lights, the white lights. There is something special about this light, unique in color, perhaps unique in its bold trespass of the fixed order of the heavens.

What should we make of it? (See color illustrations at the center of the book.)

Starry Night

NASA/2MASS

If you look up into a clear night sky, the kind of sky that existed when humans numbered in the few millions or less instead of the several billions or more, the sheer number of points of light can be overwhelming. You can, as here, easily pick out the Milky Way, the denser band of lights streaming from lower left to upper right that we now know is an edge-on view of the galaxy we inhabit and name after this band. If we watch for hours, we will see, perhaps against the backdrop of a tree in our line of sight, that the lights move, but they move together. If we do this night after night, however, we may come to realize that a small handful of these uncountable lights move independently of the great sweep of the lights across the sky. These errant lights we call planets.

Ancient Egypt: Har Decher

The ancient Egyptians, whose royal dynasties go back to about 3100 B.C.E., were the first to record the observation that a handful of the tiny lights in the vast night sky move independently of the rest. But astronomy was not well developed by the Egyptians. One principal reason for this was that ancient Egyptian arithmetic used a cumbersome notation that made the recording of counts quite straightforward but the working of fractions extraordinarily difficult. Thus, it was hard to calculate and record astronomical information such as how high in the sky a given object was on a given night.

The ancient Egyptians did, however, invent a calendar that has influenced Western Civilization ever since. This calendar depended on the motion of the star known to us as Sirius. Sirius is part of the constellation we call Canis Major, the Great Dog, which is visible in the sky near the hunter Orion, a constellation easy to find by the three bright stars in a line making Orion's Belt. But Sirius, also called the Dog Star, is the brightest true star in the sky. Indeed, the very name "Sirius" may come from an ancient Greek word meaning "sparkling" or even "scorching."

In the course of the year, Sirius, which the Egyptians called Sothis, rises

| Canis Major | Orion |

T. Credner and S. Kohle, Allthesky.com

at different times in relation to the Sun, which is the heavenly light that the Egyptians took to be the greatest and primordial god, the father of creation, called Amun Ra. Egyptian priests noticed that the first time in the year when Sothis could be seen to rise just before sunrise (what astronomers call a *heliacal rising*) was a day that usually closely preceded the annual flooding of the Nile River. This flooding was the most crucial natural event in the collective lives of the ancient Egyptians because it washed away their fields from one year's planting and simultaneously deposited on them rich topsoil from upriver for the next. The Nile took and gave life. Sothis came to be thought of not only as the herald but the cause of this annual fertilization, an embodiment of Isis, the primordial mother god. Her appearance marked the beginning of the Egyptian year.

The Sothis calendar had 365 days because this heliacal rising accurately marks—unbeknownst to the ancient Egyptians—the revolution of the Earth around the sun. To keep track of the days between the heliacal risings of Sothis, the Egyptians turned to their constellations. They had identified thirty-six constellations fairly evenly spread across the annual round of the sky. Their priests counted ten days from the rising of each constellation, thus dividing the year into *decans*, groupings of ten. Although they thus missed five and a quarter days from the true solar year, they could readjust the calendar annually by watching for Sothis. This counting of the thirty-six decans in the circle of the sky became the 360 degrees by which we still divide a circle.

Astronomy mattered to the ancient Egyptians not only because of the Nile floodings and the connection between such events and the lives of the people, but because, like other ancient peoples, Egyptians saw gods in their sky. But Egyptian gods were embodied not only in the fixed stars and in the brighter objects such as the sun, but also in animals and other natural phenomena, such as thunder. Ra himself was also embodied in the Pharaoh, a human living right here on Earth. The Egyptians seem to have been the first to notice that while the stars maintained a constant relation to each other, seven other lights moved among them: the two great lights of the Sun and the Moon, and five others, the lights we now call Mercury, Venus, Mars, Jupiter, and Saturn. The light we call Mars, though, weaker by far in intensity than Sothis and inconstant in its movements as it was, was not very important to them. The light we call Mars Egyptian astronomers called simply Har Decher, the Red One.

The Idealized Solar System

Sun   Mercury Venus Earth Mars   Jupiter   Saturn   Uranus Neptune Pluto

NASA/JPL

To most of us moderns, the familiar planets exist as part of a solar system, each planet rotating on its own axis and revolving along its own path around the massive central star known as "Sol," the Sun. In our school-child imagination, the planets line up for a cosmic dance, the distances between those paths growing exponentially as each is more distant from Sol, the whole assemblage hung in the empty vacuum of black space. But we must recognize that in fact the planets never line up like this, no human has ever directly seen their alignment from above at all, and the belief in the emptiness of the space the planets occupy is an idea less than four hundred years old. This model of the solar system is an idealization dependent on and distorting our modern scientific understanding. This was never what planets were.

What Was A Planet?

$$\pi\lambda\acute{\alpha}\upsilon\eta\tau\epsilon\varsigma \; \acute{\alpha}\sigma\tau\acute{\epsilon}\rho\epsilon\varsigma$$
planetes asteres

Our word "planet" comes from the ancient Greek *planetes asteres*, which means "wandering stars," a concept that is, at its root, a contradiction in terms.

Our word "star" comes from the Greek *aster*. We see this root in the word "astronomy," the study of the stars, and in the star-shaped flower called "aster." We also see it in "disaster." "Dis" was a prefix that indicated something negative, bad, or unlucky, as in "dis-ease" or "dys-functional." The stars were, to the Greeks, gods; they ruled our fates. To be under the sway of an evil star was to have "disaster" befall you.

The English word "star," like its Greek root *aster*, comes from an even older Indo-European root, *sta*. This root we see in such other words as "stay," "stand," and "stationary." *Sta* indicates something unmoving, fixed, like the stars in relation to each other and our fates as they are uttered by the gods.

"Fate" comes from the Latin *fatum*, which means "that which is spoken." (To be without speech is to be an "infant.") The Romans used the word "fatum" to express (in the words of the *Oxford English Dictionary*) "The principle, power, or agency by which . . . all events, or some events in particular, are unalterably predetermined from eternity." The sentence of the gods could not be changed. It was fixed, stationary, like the stars in the heavens.

But planets were not fixed. They wandered among the true (in the sense of stationary) stars. The word "planet" comes from the Indo-European *pela*, meaning spread out, broad, flat, or smooth. This root comes down to us in the English "pelagic," meaning "of the sea" (through the Greek *pelagos*, "the sea," so called for its flatness), and in "plank," and "platform" (of a spread-out shape), and "plan" (as in a floor plan or battle plan, something spread out flat to serve as a guide). The Greek verb *planein*, interestingly enough, does not mean to help follow a plan but rather to lead astray. In its passive use (as in "he was led astray"), it means to wander. A *planete* was a wanderer, but our word "planet" does not come from that noun but rather from the adjective, *planetes asteres*, wandering stars. Since true "asters" are fixed, the words *planetes asteres* should mean a class of things that cannot exist, but exist they do.

Mesopotamia: Nergal

Detail of Ancient, Old Babylonian, Sculpture, Plaque *Relief plaque depicting Nergal, God of the Underworld* 2nd millennium B.C.—Terracotta; h 14.9 cm; w 8.0 cm; d. 3.4 cm. Princeton University Art Museum. Gift of Selim Dere. Photo credit: Bruce M. White

Mesopotamia in Greek means "in the midst of rivers." (Think of "hippopotamus," "horse of the river.") The ancient Mesopotamians were those peoples whose lands centered on the area between the Tigris and Euphrates Rivers, that is, lands that today are primarily in Iraq. Starting shortly after the Egyptian discovery of wandering stars, Mesopotamians began developing their own astronomy, and brought it to the most advanced state in the pre-Christian world. By 747 B.C.E., the Babylonians in particular, one of the most important of the Mesopotamian peoples, had developed a sufficiently accurate, long, and detailed astronomical record, and accompanying mathematical ability, that they could use to predict not only lunar eclipses, which is comparatively easy, but solar eclipses as well. Of course the Mesopotamians were fully aware of the wandering stars.

As early as 3000 B.C.E., Mesopotamians associated the Red Planet with Nergal, their god of war. At that time, however, war was not a dominating concern for the people. The fertility of their lands and rivers, and their convenient geographic situation, gave them such agricultural and commercial richness that they were easily the reigning force in their part of the world, fighting mostly among themselves as one Mesopotamian empire supplanted or jostled another with readjustments of political boundaries but with relative stability for most of the peoples. Nergal, in the third millennium B.C.E., was sometimes a patron of hunting, always the lord of the underworld (that is, the land of the dead), and he who led the gods into battle. He was a war god then, but not a primary god. And for millennia, the Mesopotamians flourished.

Nergal was most revered in Nineveh, which was at one point the most populous and wealthy city in the region. This is the city that Jonah rails against in the Bible. Nineveh is gone today but once stood on the banks of the Tigris opposite the site of the present-day city of Mosul. On the land that was Nineveh, the Nergal gate has been reconstructed.

In the King James version of the Bible, the Book of Nahum begins "The burden of Nineveh. The book of the vision of Nahum the Elkoshite." Bible scholars have not located Elkosh, but they have discovered the historical events surrounding Nahum's prophecy, an event made most clear in Nahum 3:13: "Behold, thy people in the midst of thee are women: the gates of thy land shall be set wide open unto thine enemies: the fire shall devour thy bars." In this case, "bars" refers to city gates. Nahum declaims "Woe to the bloody city!" (Nahum 3:1). The city could be called bloody if it were involved in warfare, but Nahum suggests Nineveh's primary sins involve commerce, robbery, deceit, and debauchery. Despite this neglect of war as a sin of Nineveh, since the city was under the protection of Nergal, the Red Planet and god of war, it could justly be "the bloody city" symbolized by the Nergal Gate.

Scholars universally agree that the prophecy of Nahum refers to the destruction of Nineveh in 612 B.C.E. as the culminating campaign in a war against the Assyrians, whose capital was Nineveh, by the Babylonians and Medes. In the course of this battle, the Tigris River flooded (". . . with an overrunning flood [God] will make an utter end of the place" [Nahum 1:8]), aiding in the partial demolition of the Nergal Gate, the penetration of the enemy into Nineveh, and the devastation of the city by fire.

Nahum is a short form of Nahumiah, which means "Yahweh comforts." The scholarly consensus is that the main point of the Book of Nahum is to demonstrate the activity of Yahweh, the Lord, the god of the Jews, in history. It is interesting that the Jewish prophet understood that this intervention was fully significant even in deciding the fates of non-Jewish peoples. Yahweh, using the Babylonians as his instruments, was mightier than the Assyrian war god Nergal.

The wording of the King James Bible ("thy people in the midst of thee are women"), published in 1611 C.E., may convey somewhat uncertain

Department of Defense, Defense Visual Information Center

meaning to us since it does not conform to modern English grammar. The translation of this same verse in the Revised English Bible, published in 1881 (New Testament), 1885 (Old Testament, which includes the Book of Nahum), and 1895 (Apocrypha), helps us: "Your troops behave like women. The gates of your country stand open to the enemy; fire has consumed the barred gates." Here the debauchery theme seems clearer. The soldiers of Nergal "behave like women" while the attacking troops penetrate the city. The very notion of "city" is female, as in the word *metropolis*, from the Greek for "mother city." For the Assyrian troops to behave like women is to be open, to succumb to penetration. The association of warfare with rape is established early in human history, and when one masculine god defeats another, it may well be because the other acts "like a woman." The debauchery within Nineveh is punished with the symbolic debauchery of the city, the rape of Nergal's city, and the forcible throwing open of its gate (see photo). The fires of passion are used, too, as punishment in the immolation of the city. The Book of Nahum, then, suggests an intersection of the peoples and gods of several religions in the history of Nergal—Mars—in the lands in which modern astronomy was born.

Ancient Greece: Ares

Scala/Art Resource, NY

The ancient Greeks identified the red planet with Ares, their god of war, seen here in a statue. Ares has placed his shield to one side, displaying the calmly powerful, casually thoughtful demeanor so characteristic of the Greek gods. Still, he holds a sprawling infant beneath his foot.

As one might have guessed, Ares was not originally one of the major gods of the Greeks. The divine balance of intellect and athleticism seems far removed from the ferocity of war.

Ares was the son of Zeus, who was the greatest of the Olympian gods, and Zeus's wife Hera, an important goddess in her own right. Zeus had many offspring by many females—some divine, some human. According to one telling, Aphrodite, the goddess of love, arose spontaneously from the sea. (Aphrodite means "called from the foam.") According to another, she is the child of Zeus and Dione. In that telling, of course, she is a paternal half-sister of Ares.

Zeus lusted for Aphrodite, but she refused him. As a punishment, she was given in marriage to Hephaestus, the ugly blacksmith god, to whom she was notoriously faithless. One of her illicit lovers was Ares, their relationship brought to light by Apollo, the sun god. Apollo's twin sister was Artemis, goddess of the moon and of hunting. Apollo was also the god of the fine arts, medicine, and eloquence. As children of Zeus and Leto, Apollo and Artemis, like Aphrodite, were half-siblings of Ares. In Apollo's revelation of the affair of Ares and Aphrodite, and in his close and supportive relationship with Artemis, we see how the Greeks understood the relationship between violence and learning. There is a place for violence (as in hunting) and it has its appeal (kindling sexual passion) but its victories are only temporary; ultimately learning will prevail.

The dark union of Ares and Aphrodite, of war and love, yielded many offspring, including most famously Deimos and Phobos, often translated as Terror and Fear, although two other offspring worth mentioning are Eros and Anteros, loosely sexual attraction and counter-sexual attraction, lover and loved. The animals most associated with Ares were the dog and the vulture, battlefield scavengers that live on the fallen.

Although Ares was not nearly so important as Zeus himself or even Apollo or Aphrodite, he could have relations with them all and was, by the time of Homer (about the ninth century B.C.E.), firmly established as one of the twelve Olympian gods. Nonetheless, he clearly was neither liked nor admired, as is made clear in Homer's *Iliad*, the classic narrative of the war of the Greeks against Troy, also known as Ilium, a city-state that flourished in what is now northwestern Turkey. Here is a synopsis from Harold Bloom's *Homer's Iliad*:

> As the fighting continues [in Book V of *The Iliad*], the gods themselves take part: The guileful Athena [a goddess of wisdom and war who favors Athens and the other Greek city-states arrayed against Troy] even leads "manslaughtering, blood-stained" Ares, the god of war, out of the battle so he cannot help the Trojans and then herself returns to help the Achaians [the Greeks]. When the Trojan hero Aeneas, son of Aphrodite, is struck, Aphrodite shields him with her white robe but then is herself speared through the hand by the Achaian Diomedes. Shrieking, she drops her son, whom Apollo catches. The gods temporarily withdraw, but not for long as Ares soon returns to drive the Trojans on, which again incites Athena and Hera to join the battle until Athena herself helps Diomedes drive a spear into Ares' belly. The wounded Ares goes to Zeus to complain, and Zeus, calling him "the most hateful of all gods," nevertheless has him healed. Hera and Athena return to Olympus, content that they have stopped Ares.

As one might suppose, Ares, the war god embodied in the Red Planet, was a major force in the narratives of the ancient Greeks but not a figure of widespread worship.

Library of Congress, U.S.
Capitol Exhibit

Ancient Rome: Mars

Ancient Rome in many ways adopted the religion of ancient Greece, with most of the major gods taking new names while participating in the old stories. Zeus became Jupiter; Hera, Juno; Aphrodite, Venus; Hephaestus, Vulcan; Apollo, Sol (as in our solar system); Artemis, Diana; Deimos and Phobos, Fuga and Timor; and Ares, Mars. Like the Greeks, the Romans identified the war god with the Red Planet. Unlike the Greeks, however, the Romans were devoted to Mars.

Originally, Mars was a fertility god, associated with the spring. In this, he is reminiscent not of the Egyptian Har Decher, the Red One, but of the Egyptian Sothis who floods the Nile. In the oldest Roman calendar, the year began in March, the month dedicated to Mars.

Before the introduction in 46 B.C.E. of the Julian calendar by the Emperor Julius Caesar, the Roman calendar, like the Greek, was based on the Babylonian lunar calendar. However, unlike the Babylonians, but like the Greeks, the Romans used an inconvenient mathematical notation and were not particularly able astronomers. The pre-Julian Romans counted the year as beginning at the first new moon after the vernal (spring) equinox, one of two days in the year when daylight and darkness are balanced. You can determine this day, if you are careful and patient, by direct observation. Thereafter, the Romans counted ten lunations, the periods from one new moon to the next. At the end of that time, they just stopped counting and waited for the next vernal equinox. The off-calendar days are called collectively an intercalary period.

(Our names for the last months of our year reflect the old Roman calendrical practice: *Sept*ember means seventh month; *Oct*ober, eighth month; *Nov*ember, ninth month; and *Dec*ember, tenth month. We use these inherited names even though to us these months are respectively the ninth, tenth, eleventh, and twelfth of the year. Julius Caesar shifted New Year's from March to January, a month that didn't even exist for the first Romans.)

As the Babylonians knew, there is no whole number of lunations that equals the true solar year, the cosmic revolution that controls the intervals between vernal equinoxes. Therefore, merely counting lunations would soon have brought the calendar out of synchronization with the actual cycle of the seasons, with perilous consequences. For instance, if one followed such an uncorrected lunar calendar, eventually one would have been worshipping fertility gods in the fall instead of the spring. The Roman solution, like that of the Hebrews, who also used a lunar calendar and an inconvenient mathematical notation, was to add an extra month from time to time. That is, the Romans had a year of ten lunar months with two or three intercalary months, the question of two or three being decided as best they could. Despite this attempted correction, the Roman calendar inevitably became desynchronized with the solar year. The Julian calendar, which is very close to the one we use, fixed that.

Before the Julian calendar, some reset mechanism was necessary to keep the lunar calendar from drifting too far from the solar year. The Egyptians used the heliacal rising of Sothis, a sign of fertility, as the reset. The Babylonians used the heliacal rising of a bright constellation we call the Pleiades, a visible resurrection after forty days of the constellation remaining below the horizon, an event that occurs in our modern calendar about the time of Easter, also a moment of springtime fertility. The Greeks, too, understood that the solar year, not lunar counting, controlled seasonal events, but they did not use a handy stellar reset, although they could have, because, like the Babylonians (and, for that matter, the Incas) they recognized the agricultural importance of the Pleiades.

In *Works and Days* by the Greek Hesiod, who lived about 700 B.C.E., we find this advice:

Start reaping when the Pleiades rise, daughters of Atlas,
and begin to plow when they set.

For forty days and forty nights they lie hidden,
but as the year moves on in its cycle
they can be seen again when you first sharpen your iron.

(translation by Apostolos N. Athanassakis, lines 383–387)

The Pleiades, a constellation with seven stars, is also known as The Seven Sisters. The parents of these mythological figures were Atlas and Pleione, so "daughters of Atlas" refers to the stars themselves rather than to the farmers the poem advises. While we cannot make inferences about the sex of the likely reapers from this passage, we can infer that the Greeks understood that agricultural rhythms were ruled by the stars, not the moon. Individual fertile women, however, for obvious biological reasons, were more likely to be thought of as linked to the moon. This knowledge, which the Greeks shared with the other ancient peoples, becomes important later in the development of astrology: the fixed stars control the large context within which we all work and live ("works and days"), but there are other astronomical variables, like the moon and the planets, that change much more rapidly than the stars and they control us, too.

We should also notice that the Greeks associated the heliacal rising of the Pleiades with the sharpening of iron. This was for reaping in Hesiod, but for the Romans it was also for war.

While the pre-Roman Mars was only an agricultural deity, the Romans, who were famously successful conquerors, associated him with the Greek war god Ares. Springtime, when the calendar started up again and the weather improved and food could be found growing wild, was not only a good time for farmers to work but a good time for armies to move. The Romans worshipped Mars not only for the fertility of their fields but for the success of their campaigns. Their war cry was *Mars vigila!* Mars awake! According to Roman mythology, Mars was not only the son of Jupiter but the father of Romulus and Remus, the mythical twin babies who were suckled by a she-wolf and grew up to found Rome. (Just as the Greek Ares had the dog and the vulture as animal familiars, so the Roman Mars had the wolf and the woodpecker. The attack of the woodpecker on trees and palisade fortresses is in some ways reminiscent of the Babylonian attack on the Nergal Gate.) Clearly the Romans put warfare, and hence Mars, much more directly at the center of their lives than did the Greeks.

The story of Aeneas gives us one measure of this difference between the Romans and the Greeks. Aeneas, as we learn in *The Iliad*, was a Trojan, a pious man, but on the losing side for all that. By the time Virgil wrote *The Aeneid* (19 B.C.E.), the great epic of the origins of imperial Rome, there were

Re'union des Muse'es Nationaux/Art Resource, NY

many stories about how Aeneas had escaped from defeated Troy carrying his aged father on his back and holding his young son by the hand. This masculine, intergenerational trio, led by the one most capable of fighting, wandered the Mediterranean world and had, according to diverse tellings, diverse adventures until they finally came ashore at Lavinium.

In writing his epic, Virgil wanted to authorize and validate Imperial

Rome, to connect it with the gods and with the Greek civilization which the Romans admired yet had supplanted. His poem aimed to establish Rome as the inheritor of Greek culture yet different from, a superior outgrowth of, that same culture. Aeneas fit Virgil's purposes perfectly. Historical timing and cultural devotion made it impossible to try to suggest that Aeneas rather than Romulus had founded the town of Rome, but it was possible to have Aeneas found the town of Lavinium. This town in fact became the chief member of the Latin League and was crucial in changing Rome from a town into an empire. Aeneas, as Homer tells us, was protected by Aphrodite (Venus to the Romans) and yet he was also a warrior and would-be king of the Trojans, championed by Ares (Mars to the Romans). While to the Greeks the union of war and love, Mars and Venus (see photo in this chapter), led to fear and terror, to the Romans it represented an ideal. In the person of the pious Aeneas, favored by the sometime lovers Mars and his half-sister Venus, this ideal of soldierly passion became a guiding philosophy. In the Roman world, Mars was worshipped widely and enthusiastically.

The Romans called themselves the sons of Mars. The Latin word *martial* still indicates the demeanor of a noble soldier. The symbol of Mars was the lance and his element was iron, the metal of the lance (as well as of the scythes Hesiod speaks of sharpening). For any nation with interests in expansion, the Roman Mars became a key figure. The statue of Mars heading this chapter shows a much more military figure than does the Greek statue of Ares. The shield of Mars still rests to one side, but is held at the ready, as is the iron sword. The helmet is crested, like the woodpecker and like many aggressive male animals. This statue, by Luigi Persico, was completed in 1831 and is called *War as Classical Figure of Mars.* It is part of the U.S. Capitol. The significance of Roman Mars did not end with the Roman Empire.

The Sunset of Mars

Philippe Jacqueroux, www.digimond.org

The Roman Empire changed from within and without. A major internal force was Christianity, a religion that began among the marginalized peoples and social classes of the Empire but eventually supplanted the old gods. Starting about 312 C.E., at the behest of the Emperor Constantine, Christianity began its relatively quick march to becoming the official religion of Rome. In achieving this, Christianity bested not only the old polytheistic religion native to the land of Rome but another, competing religion, Mithraism, a religion that arose in Persia.

Mithraism was a vigorous force in the Roman world. Just as Christianity centered on Jesus Christ, Mithraism centered on Mithra. Mithra was the sun-god and taken to be preeminent among the gods. He was also taken to be the special patron of the ruler. Thus Mithraism was a convenient candidate for official status. Mithraism, with its complex rituals and learning, had the appeal of a mystery religion, those religions that involve an elite of the initiated, and it had the appeal of a certain kind of clarity produced by its dualistic view of good and evil. These characteristics accorded well with both the structured society that was Rome and the us-versus-them mentality that guided Roman politics. Christianity, instead of mystery, offered a faith available to all; and instead of balanced good and evil, it offered an overriding divine goodness that would, in this world or the next, put evil in its place. Perhaps most importantly, while Mithraism accepted full participation only by men, Christianity accepted men and women. Constantine either saw Christianity as the true light or as a means of binding a fragmenting empire or both. Whatever his motives, Rome became Christian.

The god of a monotheistic religion certainly can support war. "So Joshua took the whole land, according to all that the Lord said unto Moses; and Joshua gave it for an inheritance unto Israel according to their divisions by their tribes. And the land rested from war" (Joshua 11:23). Holy warriors in the name of Allah swept across the Mediterranean world from one end to another, establishing caliphates from Byzantium to Granada and beyond. And medieval Christians sought through repeated Crusades to reclaim Jerusalem from the Muslims, and sometimes they succeeded. But in all these campaigns, war was but a means to an end, to fulfilling the covenant of a homeland or to spreading the word or to controlling worshipful access to sacred territory. In no case was the god in question defined by war. Once monotheism dominated Western culture, there was no place for a war god. Culturally speaking, Mars set.

Changes in Venus

Vatican Museum

Venus, shown in this statue from the Vatican Museum with Cupid, her son by Mercury, was prominent before the rise of the war god and before the pre-Romans accommodated themselves to the religion of Greece. Beginning about the middle of the third millennium B.C.E., however, throughout the Mediterranean world, the male gods began growing ever more dominant and the female gods ever more subordinate. Nonetheless, the crucial, central, indispensable power of fertility rested with females. The brightest planet in our sky is Venus, which we now know is also both the Morning Star and the Evening Star, usually, when visible at all, the first and last heavenly object visible besides the Moon. Venus is the companion of the Moon, associated then with the periods women and the Moon share. Even when goddesses were being subordinated, these powers had to be acknowledged.

Despite the importance of God the Father, with the ascendancy of God the Son, the centrality of the mother, without whom there would be no son, was established. The relative significance of Mary in Christianity varies across the centuries and nations, sometimes taken to be only a blessed vessel and other times so revered a figure of worship as to be all but a god herself. Of course, as a monotheism, Christianity cannot consider Mary a true god or allow any of the old gods to survive unchanged. Still, in medieval iconography, the system of graphic symbols used in religious paintings to communicate to the largely unlettered congregations, Mary is occasionally shown with the Moon, almost as a companion of the Moon, rehearsing visually the role of Venus.

The more common Christian role for the planet Venus, however, has a much more infernal meaning. The word *Satan* in Hebrew means adversary, and Satan is used in the Bible's Book of Job as the designation of the powerful character who argues to Yahweh that humanity is not truly worshipful. God lets Satan test humanity as represented by Job. Job endures all and is ultimately rewarded. Satan's tests, then, may well strengthen us. While Judaism typically looked on misfortune as divine retribution, as in The Flood we associate with Noah and the destruction of Sodom and Gomorrah, Christianity often saw these punishments as chastisements, instruments, tests that either destroyed us or purified us. This harsh means of possible improvement was called "a scourge of God," a scourge being literally a whip or lash. Timur (1336–1405), the Central Asian Islamic warlord remembered in the Christian West chiefly for his cruelty in Russia, India, and the Mediterranean world, was also known as Timur the Lame, Timur Lenk, Tamerlane. In the late Middle Ages, Tamerlane was also called The Scourge of God. Those whom he barbarized doubtless had sinned, Christian priests argued. Christendom would not be able to resist Tamerlane until Christians themselves reformed. In the torments he dispensed, Tamerlane was like Job's Satan, flaying an enemy and for good measure killing his family.

Christianity came to associate Satan with Lucifer, a name meaning light bearer. In Milton's *Paradise Lost*, we read that Lucifer was the brightest of all the angels in heaven, yet Lucifer led a revolt against God and was cast out and down. He became the Lord of Pandemonium (the place of all demons), the keeper of Hell. As such, of course, he still does God's work, just as Satan does in the Book of Job. This association of Satan with rebellious Lucifer rests in part on the Book of Isaiah (14:12): "How art thou fallen from heaven, O Lucifer, son of the morning! How art thou cut down to the ground, which didst weaken the nations." In weakening the nations, Lucifer had been testing them, at least according to Christian interpretation. Thus, Lucifer was that same divine scourge Jesus speaks of in the Gospel of Saint Luke (10:18): "I beheld Satan as lightning falling from heaven."

Lucifer was another name for Venus in its role as the Morning Star. For Christians, then, these verses support the association of Venus with "the scourge of God." In Luke, though, note the word "as." Venus in Christian writing is taken to be a rhetorical representation of Satan, not a literal one. Similarly, while the planet Venus once had been the literal god of love, to Christians it was at most a symbol associated with maternal love. The logic of monotheism obliterates all gods but one. Venus hung on somehow in her figurative way, but her sometime consort, Mars, at least to Christian orthodoxy, did not. Officially, Christianity banned the gods from the skies.

Metaphorical Mars

North Wind Picture Archives

Although Christianity banned the old gods from the skies, they never truly lost their seat in the imagination. As one case in point, take William Shakespeare's history play *Richard II* (1595). Shakespeare creates for John of Gaunt (1340–1399), pictured here, what has become perhaps the most famous praise ever given to England. John, as Shakespeare's audience would have known, had been Duke of Lancaster, a son of King Edward III and father of Henry Bolingbroke, who, through many battles, became King Henry IV, and thus grandfather of King Henry V, who in turn, by his visible piety and outstanding generalship (especially in the decisive victory over France at the Battle of Agincourt, 1415), served as a model monarch in the popular imagination. John, in other words, could be viewed as a fulcrum figure moving England into a centuries-long period of dynastic and international warfare that finally led to the relative domestic peace and international glory of the England of Queen Elizabeth (1533–1603). As John lays dying early in Shakespeare's play, he prays that his death, which he knows presages conflict, will lead to victories that will secure his nation as he sees it:

This royal throne of kings, this sceptred isle,
This earth of majesty, this seat of Mars,
This other Eden, demi-paradise,
This fortress built by Nature for herself
Against infection and the hand of war,
This happy breed of men, this little world,
This precious stone set in the silver sea,
Which serves it in the office of a wall,
Or as a moat defensive to a house,
Against the envy of less happier lands;
This blessed plot, this earth, this realm, this England . . .

(II. i. 40–50)

In mixing the Roman Mars and the Christian Eden freely in his praise, John speaks of a practical realm, even though he borrows religious terms. In one sense, he replaces both religious traditions by elevating the material world he inhabits ("This fortress built by Nature for herself"). The island situation of England grants it tactical advantages here in Nature's world ("this earth, this realm, this England"). And yet, these advantages demonstrate that England is "blessed." In Shakespeare's world, religion, no matter how often sidestepped, was very much a practical matter. Elizabeth's father, King Henry VIII, had broken with the Church in Rome to secure his own dynastic succession. In so doing, he created the Church of England with the monarch as its hereditary head. If Shakespeare's conjunction of metaphors—Roman, Christian, Natural—was acceptable in his time, that was because in some metaphorical sense, though never officially in a religious sense, the old gods persisted.

Ptolemy

The popular power of the old gods flows in large part from an alliance, more than a millennium old, with the complex science of Ptolemy. While Christianity grew and eventually became institutionalized in the Roman world, science, too, by fits and starts, advanced. From the standpoint of astronomy, the greatest monument of the Roman Empire was the work of an Egyptian Greek named Ptolemy who worked in Alexandria and probably lived from about 100 to about 170 C.E. His greatest work, *The Almagest*, in all likelihood represented a culmination of the work of many others rather than a completely original contribution, yet by its brilliant mathematics and marvelous synthesis of the knowledge and observations gleaned from others, it defined how his civilization saw itself in creation. His view of the heavens, which we still call the Ptolemaic system, dominated Western thought for at least a millennium and a half.

Ptolemy argued convincingly that all heavenly motions were predictable. More than this, he argued for an Earth-centered universe in which the fixed stars exist on an outermost sphere and against which one could observe the motion of all the others (which to Ptolemy included the Sun and Moon but not, of course, the Earth) each riding a circle of its own. All motions were uniform and circular and all circles had their centers at or near the center of the Earth. Ptolemy was able to put the circles, on which these bodies rode, in order as they got further from the Earth (Moon, Mercury, Venus, the Sun, Mars, Jupiter, and Saturn). And, most astonishingly, he was able to calculate where they would be.

Of course, since this scheme does not represent reality, it would be amazing if it worked perfectly. It doesn't. But, given the accuracy of observation then available, it came close, and the geometric elegance of a universe made of circles formed a picture too appealing to reject. The cycles of Ptolemy, as observational objections were raised, required adjustments by the postulation of other cycles to ride along the primary cycles, and as the system of epicycles upon epicycles became more complex, the initial elegance of the system faded. Still, no better one was available, so Ptolemaic thought continued its long rule. There was no need for gods in Ptolemy's heavens. The stars, including the wandering ones, formed a beautiful mechanism that justified itself. And this mechanism, according to Ptolemy, ruled the lives of us all.

The Observation of the Planets

The Ptolemaic system, which saw the universe as a series of nearly concentric circles, could be represented by a disk-shaped instrument with which the relative movements of these circles could be modeled. As the theoretical circular motions had to be adjusted to conform with knowledge gained through improved observation, it became convenient to represent this more complicated system in three dimensions. Although both two- and three-dimensional devices may be called astrolabes, the three-dimensional devices, which began to appear about the 6th century C.E., are more specifically called "armillary spheres" or "armillas."

The armilla, like a slide rule, is useful for the concrete physical solution of practical mathematical problems. In this case, the practical problems were of two sorts. The first had to do with geographic location, a matter especially important for sailors. Navigators out of sight of landmarks have always sought ways to rely on the stars. The second had to do with astrology.

Astrology is the study of the influences of the stars (by which the ancients meant also what we would call planets) on our lives. Many peoples believe in heavenly omens. The Babylonians, for example, saw the gods at work in the appearance of comets, and Christians speak of the star that marked the birth of Jesus. But omens—spontaneous, irregular occurrences—are not the same as regular movements. Ptolemy used the mathematics of his great astronomical work, *The Almagest*, to develop a mathematical, predictive astrology. While today many people see a belief in astrology as the exact opposite of a belief in astronomy, Ptolemy thought of astrology, like astronomy, as a science, although a less exact one. Ptolemaic astrology was not nearly so concerned with odd events, like the passage of a falling star, as it was with predictable events, like the relation of the planets to the zodiacal constellations at the hour of one's birth. To calculate exactly which "stars" influenced a person, one needed biographical information which could be turned into astrological information by applying the tables and by using an astrolabe or armilla. Ptolemy's astrology is contained in a four-volume work known as the *Tetrabiblos*. Although in the second century c.e. the rise of Christianity was banishing the gods from the heavens, the work of Ptolemy made the observation of the planets ever more accurate and useful and their influence in our lives all the more clear and inevitable.

Astrological Symbols

Signs		Aspects	
♈	Aries	☌	Conjunctions
♉	Taurus	☍	Oppositions
♊	Gemini	△	Trines
♋	Cancer	□	Squares
♌	Leo	⋎	Semi Sextiles
♍	Virgo	✳	Sextiles
♎	Libra	⊼	Inconjuncts
♏	Scorpio	∠	Semi Squares
♐	Sagittarius	♕	Sesquiquadrates
♑	Capricorn	**Q**	Quintiles
♒	Aquarius	⑦	Septiles
♓	Pisces	▷ ◀	Noviles
		//	Parallel

Astrology has its roots in the priestly observations of the heavens, activities common to many ancient peoples, including the Egyptians, Babylonians, Greeks, and Romans. In the Christianized Roman Empire, dominated by Ptolemaic astronomy, and on into the Middle Ages, astrology developed into a complicated, detailed system. That system, which has continually adapted to changes in the astronomical understanding of the universe, remains with us. In fact, in the United States today, there are about twenty professional astrologers for every one professional astronomer.

In keeping track of heavenly bodies and their relations to each other, astrologers use a series of time-honored symbols. The signs represent the twelve constellations of the zodiac, each rising after another, counting out a zodiacal year corresponding to a post-Julian calendrical year. These signs, of course, remain constant in their motions and relations to each other. They track those stars that are fixed to the empyrean, the outermost sphere of the heavens.

The planets, the wandering stars, move about in relation to the zodiac. The table here, which includes Uranus, Neptune, and Pluto, obviously pertains to the work of modern astrologers since those planets were unknown

Planets					
☉	Sun	☽	Moon	☿	Mercury
♀	Venus	♂	Mars	♃	Jupiter
♄	Saturn	♅	Uranus	♆	Neptune
♇	Pluto	☊	North Node	☋	South Node
⚷	Chiron	⚴	Pallas	⚵	Juno
⚳	Ceres	⚶	Vesta	⊕	Earth

to Ptolemy. (Pluto, in fact, was discovered only in 1930.) However, not all the planets here are what one might expect. Although Chiron is the name of a body (some say an asteroid, some say a comet, some both) in the asteroid belt that runs between Mars and Jupiter, Chiron was discovered only in 1977. The Chiron symbol from this table has nothing to do with that body. Rather, it is placed in the birth chart of an individual to reflect certain archetypal energies associated with shamanic powers and accidental wounding. (In Greek mythology, Chiron was a gentle, wise, learned centaur—half-man and half-horse.) This ambiguous condition—powerful yet wounded—is something a skilled astrologer will know how to add to one's chart just as the astrologer will know how to place the other planets that reflect the influences on and characteristics of the person, some of which go back to Aristotle's beliefs about the fundamental powers of hot and cold, dry and wet. But the heavenly planets will be placed by use of astronomical tables, following in the footsteps of Ptolemy.

The aspects capture the relations of the planets to each other and to the zodiac. Some of these aspects can be precisely calculated, such as whether or not two true planets are in conjunction. (Conjunction means coupling in biology, union in mathematics, proximity in astronomy, and, in astrology, the lining up of the objects as viewed from Earth.) Other aspects cannot be precisely calculated not because the concepts are ill-defined but because they are applied to "planets" that are placed by the art rather than the science of the astrologer.

Within this system, Mars is called the lesser malefic, that is, an evil planet (and Saturn is the great malefic). The evil potency of Mars flows from the bellicose nature of the Roman god, a deity who projects aggression, self-assertion, and ego. That is, those dominated by Mars feel the urge to distinguish themselves as unique even at the cost of those around them. Most of the world's major religions today, however, seek peace as a goal for the world and, for the individual, a unification with some greater power or with the universe as a whole. In this sense, the aggressive egotism of Mars constitutes evil.

The symbol for Mars can be thought of as a body with a lance, a fit representation of the god of war. It can also, of course, be thought of as a phallic symbol. (Such a view recalls the association of military penetration with sexual penetration, as in the destruction of the Nergal Gate when the soldiers of Nineveh were as women.) This gendered interpretation accords with the traditional use of the Mars symbol to indicate male and the Venus symbol to indicate female. The cross below the circle in the Venus symbol can be thought of as the lines made by the creases of a woman's lap and external sexual organs. Mars and Venus were mythological lovers. In astrology, they are potentially lovers still. We must watch them both—and, as astrology tells us, the rest of the heavens besides—if we are to know and thus take advantage of our fates. In a sense, the Venus and Mars of astrology participate in a shadow religion, crucial parts of the repopulation of the heavens that monotheism had scoured.

Days of the Week

English	Spanish	Roman God	Norse God
Sunday	domingo	Sol	
Monday	lunes	Luna	
Tuesday	martes	Mars	Tir
Wednesday	miércoles	Mercury	Woden
Thursday	jueves	Jupiter/Jove	Thor
Friday	viernes	Venus	Frigga
Saturday	sabado	Saturn	

The seven-day week with names we would recognize today was established in Rome by the fourth century C.E. In the table above, the days are listed in English and also in Spanish, today the most widely used of the languages that descend directly from the Latin of Rome. It is sometimes easier to see mythological and astronomical origins of the day names in Spanish than in English.

SUNDAY

The Sun was the greatest of the ancient gods and important in all polytheisms. He was the Egyptian Amun Ra and Osiris, the Greek Apollo and Helios, the Canaanite Moloch, the Persian Mithra, and so on. Sun worship continued even in Christianized Rome. Sol was the Roman sun god, and his day was the first of the Roman week. To co-opt his worship for Christianity, in 321 Emperor Constantine decreed that the day of the sun would be a day of rest. This arrangement served to shift but maintain the tradition in which the Hebrews had rested on the seventh day of the week, a ritual rest modeled on God's creation: "And on the seventh day God ended his work which he had made; and he rested on the seventh day from all his work which he had made" (Genesis 2:2). Since Sol was the lord of the pagans and Jesus the lord of the Christians, this bit of appropriation was eased by using neither name but instead calling the day of rest the lord's day. *Dominus* is lord in Latin, the root of the Spanish *domingo*. But the English Sunday is an accurate translation of the Latin day name, *diēs sōlis*, the day of Sol, before Constantine Christianized it.

MONDAY

Monday is the day of the Moon, the second of the great lights and, in pre-Christian Rome, the cool female principle that becomes inflamed by the male Sun. *Luna* is Latin for moon; hence the Spanish *lunes*. English Monday is simply a variant of Moon's day.

TUESDAY

Martes in Spanish recalls the naming of this day after the Roman war god, Mars. In the mythology of the Norse, a powerful northern European culture influencing the ancestors of those whose language would develop into English, Tir or Tyr is an early war god. As Norse mythology developed, Tir was subordinated to Odin, the leader of the gods and the god associated with battle, inspiration, and death. In this view of the Norse pantheon, Tir becomes a god of honorable war, just as Mars functions for the Romans within the limits set by Jupiter. The day of Tir was Tiw's day, a translation of the day of Mars.

WEDNESDAY

Mercury (Hermes to the Greeks), the messenger of the Gods, was, like Mars, a son of Jupiter. His name is apparent in the Spanish day name *miércoles*. The English Wednesday is not a translation. Rather, Wednesday is Woden's day, Woden being the much worshipped Anglo-Saxon equivalent of the Norse Odin.

THURSDAY

Although Sol was the greatest god of the pre-Roman pantheon, among the Romans, the leader of the gods was Jupiter or Jove. His name is apparent in the Spanish *jueves*. Jupiter had many attributes, but perhaps most palpable among them was his command of the heavens themselves. Jupiter's thunderbolts could blast from the sky to settle matters of life and death. Jupiter was the god of thunder. While the Norse Thor was not greater than Odin, Thor was the god of thunder. Hence Jupiter's day in Rome became Thor's day in northern Europe and Thursday in modern English.

FRIDAY

The sixth day of the Roman week belonged to Venus, the goddess of love. Her name remains visible in the Spanish *viernes*. Among the Norse, the wife of Odin (who would be equivalent in rank to Juno rather than to Venus) was Frigg (pronounced *Frih*) and among the Anglo-Saxons Frija (pronounced *Fri-ya*). As the consort of the greatest of the gods, the wife of Odin represented fertility. Thus the day of Venus became Frija's day, Friday in modern English.

SATURDAY

Saturn (Cronos to the Greeks) represented time itself. He was a Titan and father to six of the twelve Olympians, including Jupiter, who overthrew him. We see his name still in Saturday. The Spanish *sabado*, however, comes from the Hebrew *shabbāth*, meaning the day of rest, from the Hebrew verb *shābath*, to rest. Thus, thanks to Constantine, the seventh day of the week in the Christian world is called a day of rest while the Christian first day, Sunday, is the day of rest. When Emperors and the great gods settle a matter, even gods of war may be put in their place.

Alchemy

△	♂	♀	♃	♙
Fire	Iron	Copper	Tin	Sulphur
☉	☽	☿	♄	♂
Gold	Silver	Mercury	Lead	Antimony

In the fourth century C.E., in the same era when the days of the week insti-tutionalized the ancient gods in the languages of the Christian world and Ptolemy's view of the universe fixed Earth at the center of a mobile yet pre-dictable creation, alchemy began to grow in importance. If we consider all extant texts, we can find different "alchemies" arising first in China (c. 700 B.C.E.), then in India (c. 400 B.C.E.), then in Hellenistic Greece (c. 300 B.C.E.–c. 300 C.E.), then in the Arabic world (c. 700 C.E.), and finally in the Latin world of the Holy Roman Empire (c. 1150 C.E.).

These alchemies had somewhat different views of the material world, but they had in common two features. First, they were sibling systems of thought to astrology. As astrology was concerned with understanding the powers of the superlunary heavens, the realms at the Ptolemaic spheres above the Moon, alchemy was concerned with understanding the powers of the sublunary world. Second, these alchemies all sought to understand the elements (which for them included fire!) that made and moved the world so that the alchemist might transform one element into another. One etymology of "alchemy" traces it through the great Arabic scholars of Egypt back to a Greek word (*chymia*) used to refer to transmutation, and specifically (as in the Decree of Diocletian against the Hellenized Egyptian al-chemists, c. 300 C.E.) the transmutation of less valuable metals into silver and gold.

In alchemy, as in astrology, the planets had their place. The alchemical symbols crucial for recording the recipes and researches of the alchemists often came from astrology. Some of these correspondences make obvious sense. The circle that represents the Sun, which is a circle in the sky, stands for gold, which accords with the sun god being the most important and gold the most precious of metals. The crescent that represents the Moon, which is often a crescent in the sky and is the next most visible heavenly object after the sun, stands for silver, the next most precious metal. The Sun, of course, is colored gold and the Moon silver. And both were gods.

Some of the alchemical symbols are less than obvious. An equilateral tri-angle in astrology stands for "trines." This is the relation of two heavenly bodies one-third of the way around the zodiac from each other, that is, 120° apart in the circle of the sky. From this position, they may approach each other or retreat, a fact considered crucial in understanding our annual fates since three trines times four seasons make twelve signs of the zodiac (360°). This same triangular symbol is used for fire in alchemy because fire may fuse two materials together or may cause a compound to separate into compo-nent parts.

Within this system, Mars has its expectable place. Since Mars is the Red Planet, it becomes the god of war. Since warriors use implements of iron, the symbol of Mars becomes the symbol of iron. Iron is the most common metal that offered the experiences that drove the alchemists. From ancient times, people understood that subjecting material to fire could sometimes destroy it and sometimes improve it. Wheat could be burned or bread baked from it. Clay could be powdered or hardened into pottery. Sand could be fused into glass which, unlike sand, could hold perfume, but also unlike sand, could shatter. And iron could be softened if overheated or hardened if heated just enough and then quickly quenched. A sword could be beaten to a keen edge which fire and water then fixed, making the sword—and the iron—more valuable.

The process by which iron is fired and strengthened we called annealing. The word *anneal* originally meant to bake or set on fire. This transmutation of Mars was alchemy in action.

The alchemists never conquered our star-writ fates by finding their much-sought Elixir of Life nor did they succeed in transmuting lead into gold. But they maintained, studied, and advanced the tradition that, among other feats, improved iron. The Arabs, by their swords and faith, conquered much of the Roman Empire, bringing their learning, including their alchemy, with them. By the mid-twelfth century, the Christian prejudice against Muslim learning fell before the unarguable grandeur of those outsiders. Christians began to read Arabic alchemy, translate it into Latin, and finally join all the alchemies together. In Europe, this proto-chemistry held sway for five centuries. Its twin, astrology, came under serious attack in only three centuries.

The Mars Symbol

The symbol comprised of a circle with an upward-pointing arrow is one of the oldest we retain today. It has many uses: In mythology, it is the war god Mars; in astronomy and astrology, the planet Mars; in alchemy, the element iron for the element's rusty color and bloody military association with Mars.

Because of the importance of the god Mars, he also has a month, March, and a day, Tuesday (Latin *dies Martis*), both sharing his symbol.

The Mars symbol continued to accrue meanings, for example in the early branch of learning called physiology. Medieval physiology was not the science we know today but a system of beliefs about the human body and character that went back at least to Aristotle. According to this system, human psychology was influenced by the balance among four bodily fluids called humors (yellow bile, black bile, blood, and phlegm) and among the four elements (air, earth, fire, and water). A person who was quick to anger, for example, doubtless had too much yellow bile, also known as *choler*, a notion that left its trace in our modern word *choleric*. You could see that someone was choleric in part by a tell-tale yellowness.

Within physiology, the Mars symbol stood for blood (its redness), the external sex organs (which swell with blood like the upward-pointing arrow), the adrenal glands, muscle tissue, and the nose (which according to medieval physiologists visually mimics the external sexual organs [both penis and clitoris] and was believed to indicate one's sexual nature). We still have the phrase "a young blood," which once meant a sexually eager young man although now it means simply an energetic and hopeful young man. The modern English word "sanguine," from the Latin for blood (*sanguis*), means "of a hopeful disposition," but also can be used to mean "of a bloodred color," "of a bloodred complexion," or "sanguinary," that is, "bloody-minded" or "relating to the slaughter of war." In short, in physiology, the red, penetrating planet and god came to represent the bloody and penetrating parts of a person and a set of sometimes contradictory dispositions.

In modern anthropology, the Mars symbol is still often used in field notes to record a male.

Even before Ptolemy, the period of Mars in the sky—the time needed to move from one point against the backdrop of the zodiac, cross the field of stars, and return to that same point—a length of time nearly what we would call its year, was known to be about as long as two of our years. Thus, when modern botany needed symbols, they adopted that of Mars to indicate plants with a two-year growing cycle.

In the twentieth century, hobos and Boy Scouts adopted the circle with an arrow to indicate "go this way," although they may have been interested only in the arrow and not in Mars.

On soldiers' field maps, the symbol for a grenade thrower or a mortar is also a circle with an upward-pointing arrow. Just as the hobos' symbol mimics a pointing hand, the soldiers' symbol mimics the actual weapon. Whether soldiers originally intended it or not, the association of the mortar with the god of war is inevitable.

Nicolaus Copernicus: Reorganizing the Universe

Library of Congress

Nicolaus Copernicus (1473–1543) did not, it seems, set out to reorganize the universe. He just couldn't help himself.

Copernicus was born into a well-to-do merchant family in Poland. After the death of his father (c. 1484), his mother's brother, a man soon to become a bishop, took charge of Copernicus's upbringing. Copernicus was destined for church service and indeed became a canon. Before settling into those duties, about 1503, he had his preliminary schooling in Poland and then was educated in Italy in mathematics, astronomy, astrology, and medicine. Since astrology was the mathematical application of astronomy to the lives and affairs of people, the first three of those fields were highly allied. Since medicine in this period was concerned with humors and the influences of the heavens on our bodily composition, astrology and medicine were allied. Once Copernicus returned to Poland to work in his uncle's bishopric, he used all his diverse learning in fulfilling his administrative, household management, medical, and astrological chores. In the sense that Copernicus was vigorously involved with practical matters, astronomy was perhaps a sideline. Indeed, while Copernicus was clearly deeply interested in the theoretical bases for astronomy and astrology, his known observations number only twenty-seven.

While at the University of Bologna, Copernicus served as an assistant to Domenico Maria de Novara (1454–1504), the chief star-scholar at the university, an astronomer charged with issuing annual astrological prognostications for the city, its social classes and leading families, and its enemies. Working with Novara, Copernicus became familiar with two crucial works, one by Johann Müller (1436–1476) and one by Giovanni Pico della Mirandola (1463–1494). The former work was the authoritative, then-modern digest of Ptolemy's *Almagest* and of the whole Ptolemaic, geocentric system, including its astrological branches. The latter work was a serious attack on the very reliability of astrology.

In fact, the Ptolemaic system was far from perfect astronomically. As Pico stressed, it could not predict all astronomical events with accuracy. More importantly, it could not resolve a question upon which eminent astrologers disagreed: What was the order of the celestial circles of the planets from the Earth? If astrologers could not even agree about the planets' proximity, how could one trust astrologers' inferences about the relative strengths of the influences those planets had upon us?

Between 1508 and 1514, already serving his uncle officially, Copernicus wrote a manuscript he called "Little Commentary." Judging from its contents, historians believe this was probably the period during which Copernicus came up with his radical, heliocentric—sun-centered—view of the universe. However, the final form of his system was first published only in 1543, the year of his death, in his *De revolutionibus orbium coelestium libri VI* (Six Books Concerning the Revolutions of the Heavenly Orbs). The sidereal periods of the planets—the length of time necessary for a planet to move against the background of the stars from any given position back to that same position—were, of course, already known: Mercury (88 days), Venus (225 days), Earth (1 year), Mars (1.9 years), Jupiter (12 years), and Saturn (30 years). By postulating that the planets orbited the Sun, these periods nicely settled the question of order: the longer the period, the further from the sun.

But if the Earth, along with the other planets, circled the Sun, and the Sun did not circle the Earth, how did Copernicus explain the apparent motion of the Sun across the sky? The answer, so obvious to us, was stunning to his contemporaries: The Earth not only revolved around the Sun but rotated on its own axis.

The notion of rotation helped solve another problem, one much less troublesome to those interested in astrology but equally important to those interested in astronomy. For any object spinning on an axis, it is possible for the axis itself to rotate. One can easily picture a child's top rotating swiftly, the lower end of its axis stable on the nursery floor and the upper end of its axis tracing out a slow, steady circle in the air. That slow circling, not the rapid rotation around the axis but the rotation of the axis itself, is called precession. For any freely spinning object, precession has a fixed period. Why did this matter to Copernicus?

In 129 B.C.E., the Greek astronomer Hipparchus completed a famous star catalog. He noticed that the positions of the stars had systematically shifted from the positions assigned them by the much earlier yet expert astronomers of Babylon. That the shifts were systematic indicated that the stars had not moved in relation to each other but that the position of the observer had moved in relation to the stars. As long as one imagined the Earth as the center of the universe, this crucial observation could not be explained. Still, it did not have practical consequences because the shift was small and only observable over many centuries. However, once one postulated a rotating Earth, one could postulate a precession for the axis of rotation, a precession that would shift the Earth as a point of observation in relation to the stars. Thus Copernicus not only reorganized the universe by postulating two motions, revolution and rotation, but discovered a third, the precession of the Earth. The period of Earth's precession we now take to be about 26,000 years.

Copernicus's heliocentric system was, in theory, radical and shocking. If God created Man in his own image, if Man were the special concern and highest creation of God, then how could it be possible that Man's special home was just another rock in the sky? One might have thought that Copernicus would have come under strong attack for his views that, after all, by undercutting our received ideas of God's doings, undercut the authority of those—the Church—who foster those ideas.

In subsequent times the contest for authority between religion and science was vigorous. The 1684 engraving shows Copernicus between a crucifix, a symbol of the church which employed him, and a celestial globe, a symbol of the realm he studied both in service of the church (astrology) and for his own theoretical interests (astronomy). The Latin inscription translates to "Not grace the equal of Paul's do I ask / Nor Peter's pardon seek, but what / To a thief you granted on the wood of the cross / This I do earnestly pray." That is, Copernicus is shown, in a sense, supposedly turning his back on the study of the skies to seek his truer salvation in the grace Jesus dispensed on the cross and which he continues to dispense from the heavens.

This graphic demonstration of a clear preference for religion over science was not unreasonable. Copernicus was, after all, a church canon. But the artist may have been, by this comparatively late date, somewhat subversive. Standing beside the globe is a compass, an instrument used for measuring distances and drawing arcs on surfaces such as maps and globes. In reality,

Non parem Pauli gratiam requiro
Veniam Petri neq poſco. ſed quam
In crucis ligno dederas latroni
Sedulus oro.

NICOLAUS COPERNICUS, THORUNENSIS PRUSSUS, MATHEMATICUS CELEBERRIMUS
Ex Monumento Thorunensi expreſſus.   J. J. Vogel fecit francf.

a compass could not stand alone. However, by showing it standing nonetheless, the artist is able to make the angle of the compass' legs mimic the hands and forearms of the supposedly prayerful scholar, just as the celestial globe mimics Copernicus's head. One sees that one of the compass' legs touch the man's shoulder, from which rises the neck that supports his head, and the other of the compass' legs touches the globe's base, from which rises a spindle that supports the globe. This confluence of symbols seems to suggest that Copernicus's way to the heavens is through the compass at least as much as through prayer.

Just behind the base of the cross is a human skull. A skull often served as an object of contemplation for religious Christians from the early Middle Ages onward. Called a *memento mori*, the skull helped remind one of the inevitability of death and hence the overwhelming significance of eternal as opposed to daily concerns. Hamlet is motivated to such contemplation when he holds Yorick's skull in Shakespeare's play (1600): "To be or not to be, that is the question." But the question for us is why Copernicus is pictured here not looking at the skull at all but rather resting his left elbow on it. Yes, Nicolaus observes the cross, but perhaps the artist is suggesting, unrealized by casual observers of the etching, that Copernicus's more lasting, scientific researches paid little attention to matters of spiritual life and death.

This contest of authority became public and significant, but not for Copernicus. He was spared involvement in controversy by two factors. First, because his work was not published until the year of his death, he simply was unavailable for debate. Second, it turned out that while the Copernican system neatly solved some problems, like the order of the planets and the shift of the heavens from the Babylonians to the Greeks, it did not solve all, most notably the strange motion of the planet Mars. As long as Mars refused to conform, the Copernican system could be dismissed by those who so chose. It was left to others to refine the work, put Mars in its place, and make the clash of authority inevitable.

Johannes Kepler: Putting Mars In Its Place

Johannes Kepler (1571–1630) was in many ways more an astrologer than an astronomer. No matter how mathematical or scientific his work and research became, his motives were always spiritual, based in a belief that a living god is reflected in the heavens and directly rules our lives. The son of a poor mercenary soldier, Kepler began his studies on a Lutheran pastoral scholarship in his birthplace, Württemberg, Germany. In 1589 he entered the Stift, the theological seminary, at the University of Tübingen, from which he received an MA in 1591. At Tübingen, he studied—and accepted—Copernican astronomy from Michael Maestlin (1550–1631) to whom he later described himself as a "Lutheran astrologer."

In 1594, Kepler suspended his theological studies to become a mathematics teacher at the Lutheran school in Graz. But six years later, along with all other Lutherans, he was forced to leave Graz, at which point he turned to Tycho Brahe. Brahe (1546–1601), a Dane, was the best observational astronomer of his time, performing astonishing feats with his unaided eyes. (The telescope was not invented until about 1608, after Brahe's death.) Brahe's skill attracted patronage, including that of Frederick II, King of Denmark, who, in 1576, gave him title to the island of Ven on which Brahe established the world's first modern observatory, which he called Urania after the ancient Greek muse of astronomy. In 1599, Brahe became the official astronomer of Rudolf II, the Holy Roman Emperor, and set up in Prague. In early 1600, Kepler had assisted Brahe for a few months. After Kepler's expulsion from Graz, he sought and was granted a permanent job assisting the master, but in 1601 Brahe died. Kepler suddenly inherited Brahe's post and with it Brahe's voluminous observational records. These were to supply Kepler with the materials from which, with his imagination and mathematical skill, he built his astronomical fame.

At this period, the Copernican system was not widely accepted. Even Brahe did not believe in it. (He didn't believe in the Ptolemaic system either, for that matter—he had his own!) This failure of Copernican thought to win adherents reflects the fact that Copernicus's system had obvious flaws. Some rejected Copernicus on theoretical, religious grounds, because clearly God could not have meant Earth to be anywhere but the center of the universe. But Kepler, at least to his own mind, was able to resolve this problem with his own spiritual logic. For Kepler, the Lutheran astrologer, astronomical fame was secondary to a mathematical worship of God. Hundreds of Kepler's astrological horoscopes survive. Most impressive, although now discredited, was Kepler's use of Copernicus to confirm Divine Purpose.

Working within the Copernican system, Kepler constructed a new cosmological system. In his *Cosmographical Mystery* (1596), Kepler articulates a Christian vision of a heliocentric, Trinitarian universe with the sun as God at the center. The stars, which reflect light, represent Jesus, and the planets and space between the center and the outer sphere stand for the Holy Spirit. In Kepler's view, the divine light from the sun moves the planets in their paths.

Kepler supported this spiritual interpretation of the Copernican system with an extraordinary mathematical feat. While Ptolemaic astrologers could not settle the question of the relative distances of the planets from Earth, Copernicus easily settled the question of the relative distances of the planets from the Sun. Kepler pondered those relative distances and eventually made a remarkable discovery.

At this time, Kepler, like Copernicus, assumed that the orbits of the planets were circular since, as every spiritual geometer from the ancient Greek Pythagoras on had known, a circle is a perfect figure and therefore favored by God. There are other "perfect" geometrical figures, too, for example, the so-called regular polygons, two-dimensional shapes that have straight-line sides of equal length joined at equal angles (for example, 60° for an equilateral triangle, 90° for a square). Kepler sought to find a set of relationships among these regular polygons that would somehow reflect the relationships among the planets. But he failed.

Then Kepler tried a new idea. What if the regularity were not two-dimensional but three-dimensional? Just as there are regular polygons, there are regular polyhedrons—solids. These, like a cube, have faces that are regular polygons and have the same number of faces meet at each corner. As even ancient geometers knew, there are just five of these regular polyhedrons: the tetrahedron (four-sided pyramid with triangular faces), cube (six-sided with square faces), octahedron (eight-sided with triangular faces), dodecahedron (twelve-sided with pentagonal faces), and icosahedron (twenty-sided with triangular faces). Providentially, as far as Kepler knew, there were also just five planets. His remarkable idea was to nest one of each of these regular polyhedrons with a common center so that he could derive the relationship among their radii, that is, the closest point from their center to one of their faces. By placing them in just the right order, he was able miraculously to discover that these ratios equaled those among the relative distances of the planets, thus mathematically confirming the divine order of the universe in general and of the Copernican universe in particular.

As it happens, this remarkable coincidence is mistaken. Observations of the planets, especially after Kepler's next great discovery, did not quite tally with the idea of God the Regular Geometer. Nonetheless, Kepler published his cosmographic views, and diagrams of his discoveries, in *Cosmographical Mystery*, and maintained this view for his entire life.

History might have forgotten Kepler if he had been merely an astrologer, no matter how imaginative, but of course in those days all great astrologers were also astronomers. As such, Kepler took up the second great objection, a nonreligious objection, to Copernicus's system. The heliocentric universe failed as a model in some observational regards, by far the most important of which concerned Mars. Even before Ptolemy, the sidereal period of Mars, which is nearly what we would call its year, was known. Ancient observers could pick Mars out as a pale pink object in the predawn sky. On succeeding days it moved in relation to the stars and grew brighter as it rose earlier and earlier. Then, like the Sun reversing the direction of its march across the sky at the summer or winter solstice, Mars, by then decidedly red and the brightest object in the night sky after the Moon and Venus, reversed its march among the fixed stars. This reversal is called retrograde motion. After about seventy days of such motion, Mars stopped and reversed again, growing daily dimmer as it set earlier and earlier in the evening sky until it was again only a pale pink object this time visible just after sunset. Then it disappeared completely for about a hundred days when it suddenly appeared again, renewing its invariable cycle, each round of which spanned just about two of our years.

280   DE MOTIB. STELLÆ MARTIS

Cap.
LIX.

PROTHEOREMATA.

I.

SI intra circulum defcribatur ellipfis, tangens verticibus circulum, in punctis oppofitis; & per centrum & puncta contactuum ducatur diameter; deinde a punctis aliis circumferentiæ circuli ducantur per pendiculares in hanc diametrum: eæ omnes a circumferentia ellipfeos fecabuntur in eandem proportionem.

Ex l. 1. Apollonii Conicorum pag. xxi. demonftrat COMMANDINVS *in commentario fuper* v. *Spharoideon* ARCHIMEDIS.

Sit enim circulus A E C. *in eo ellipfis* A B C *tangens circulum in* A C. *& ducatur diameter per* A. C. *puncta contactuum, & per* H *centrum.* Deinde *ex punctis circumferentia* K. E. *defcendant perpendiculares* K L, E H, *fecta in* M. B. *a circumferentia ellipfeos. Erit ut* B H *ad* H E, *fic* M L *ad* L K. *& fic omnes alia perpendiculares.*

II.

Area ellipfis fic infcriptæ circulo, ad aream circuli, habet proportionem eandem, quam dictæ lineæ.

Vt enim B H *ad* H E, *fic area ellipfeos* A B C *ad aream circuli* A E C. *Eft quinta Spharoideon* ARCHIMEDIS.

III.

Si a certo puncto diametri educantur lineæ in fectiones ejusdem perpendicularis, cum circuli & ellipfeos circumferentia; fpacia ab iis refciffa rurfum erunt in proportione fectæ perpendicularis.

Sit N *punctum diametri, & * K M L *perpendicularis. connectantur figna* K.M. *cum* N. *Dico, ut* M L *ad* L K, *feu per.* I. *ut* B H, *ad* H E *diameter breuior ad longiorem, fic effe aream* A M N *ad* A K N. *Eft enim* A M L, *area ad* A K L *aream, ut* M L *ad* L K *per affumpta* ARCHIMEDIS *ad pr.* v. *Spharoideon, qua* COMMANDINVS *in commentariis ad hanc propofitionem literis* C. D. *demonftrat. Triangulorum vero rectangulorum* N L M, N L K, *altitudo* N L *eft eadem*

Library of Congress

Copernicus had no way to explain the so-called retrograde motion of Mars, but Kepler, who was indeed a great geometer, believed both in God and in Copernicus. Kepler broke with Ptolemy and Copernicus and postulated that the orbits of the planets were not circles with the sun at their center but ellipses with the sun at one of their two foci. If this were so, for the Earth as well as for Mars, then the retrograde motion viewed from the Earth made sense. Using the observations he had inherited from Brahe, Kepler was able to trace the path of Mars more exactly than anyone ever had before. And assuming that the path was elliptical, he was able to show that a line drawn from the focus at the sun to the planet traced out equal areas in equal times. These two ideas, that the orbits of the planets were ellipses and that they traced out equal areas in equal times, have come to be known as the first and second of Kepler's three laws of planetary motion. And, as far as we know, they are right.

Kepler made these discoveries while working in Prague from 1601 to 1605, although he first published them in *New Astronomy* in 1609. He called this intellectual effort his "war on Mars." He also contributed a third law of planetary motion in his *Harmony of the World* (1619). That law states that the square of the period of a planet's orbit is proportional to the cube of its mean radius. In that book, Kepler's point was not to discover an astronomical law but to demonstrate the musical harmony of the universe. These squarings and cubings bore the same sorts of relationships, in Kepler's mind, to those of the vibrating strings of musical instruments. Kepler never called his laws of planetary motion laws at all. To him he was merely describing the realities of "celestial music." But even he knew that that music was the triumphal march for the victor in his "war on Mars," the struggle through which Kepler, both as astrologer and astronomer, put Mars in its place.

Galileo Galilei: Questions of Authority

Galileo Galilei (1564–1642) enjoyed controversy. A brilliant scientist and a gifted writer, in matters of intellectual disagreement he relished victory at least as much as truth. He always sought both. In Galileo's world, there were two intellectual authorities above all others: The Church and The Philosopher, the term by which Aristotle (384–322 B.C.E.) was often known, just as Shakespeare later was often known as The Bard. In his lifetime, the authority Galileo first controverted was The Philosopher.

As a practicing astronomer, Galileo's importance comes more from the priority of his work than from its brilliance. Hearing in 1609 that a Dutchman (probably Lippershey) had used two lenses to make objects appear closer, Galileo grasped the principle at once and in a single feverish night built a telescope. He was the first person ever to use such an instrument for exploring the heavens. But his discoveries were largely matters of direct observation, drawing conclusions that others would doubtless have reached shortly had he not arrived first. At that time, though, he was already famous for his work as what we would call a physicist. His prior fame sped news of his new astronomical observations into the ears of many.

The beliefs of The Philosopher had constrained the development of western science. Aristotle believed, for example, in a geocentric universe, thus lending authority to the Ptolemaic position. In the sublunary realm, for example, Aristotle asserted that the rate at which objects fall is proportional to their weight. Galileo, in his famous experiment at the Leaning Tower of Pisa, demonstrated that The Philosopher was wrong, that the rate of descent has no relation to weight. This overturning of Aristotelian authority was but one of Galileo's many successes.

It was Galileo who demonstrated the regularity of the period of a pendulum, giving rise later to vastly improved timekeeping. He derived by experiment laws of mechanics, for example in relation to objects on inclined planes, that still stand. It was he who first derived the laws of ballistics. And he even came up with a system of mathematical "infinitesimals" that foreshadowed the calculus. While some "natural philosophers" who held tightly to Aristotle were angered by Galileo's incontrovertible proofs, there was little they could do. Aristotle had no practical power in Renaissance Italy. The Pope, however, did.

Galileo already believed in Copernicus when he made his telescope. Despite its failure to solve some practical problems better than did the Ptolemaic system, Galileo held to the Copernican system without wavering. It had a kind of mechanical elegance that he appreciated. Thus, he was pleased when at least one of his observations advanced the Copernican cause directly.

If Copernicus were correct, from the Earth one should be able to observe phases of Mercury and Venus just as we observe phases of the Moon. Since no such phases were visible, Copernicus had had to postulate that those planets were somehow transparent and that the light of the sun passed right through them. Those who held with The Philosopher, however, rejected this Copernican contrivance and used its patent implausibility as one reason to stick with Ptolemy.

With his telescope, Galileo observed the phases of Venus. What had been a reason to reject Copernicus suddenly became a reason to accept him.

Galileo made other observations that undercut The Philosopher. He observed moons around Jupiter, thus offering an analogue for the planets circling the sun and Luna the Earth. He looked at Luna and was able to see shadows lengthen and shorten on its surface, demonstrating that it had hills and valleys, like the Earth, and was not some perfect heavenly sphere. He found sunspots and by observing them demonstrated that even the Sun itself rotated. Others would doubtless have made the same observations soon, but Galileo made them first, he was famous and he liked controversy.

The Church at this period hardly took the trouble to attack Copernican views. Indeed, Copernicus had been a churchman and Galileo dedicated some of his own writing to the Pope, apparently with permission. The learned men of the Church understood that the Copernican and Ptolemaic systems contested for accuracy and sufficiency in explaining astronomical phenomena, and so long as that was all they did, the Church was pleased to stand to one side.

Questions of authority arose, in theory, only when one held that the Copernican view contradicted Scripture. For example, where Joshua (Joshua 10:12–14) was said to have called successfully upon the Sun to stand still in the sky, the Church needed to assert that that was a moment in which the system—Copernicus's or Ptolemy's—just didn't apply. Galileo could not accept that convenient loophole any more than he could accept convenient Copernican planetary transparency. And he said as much. Eloquently. Loudly. And often.

Galileo's ultimate attitude about the relative value of science and scripture is clear in this passage from his work called *The Assayer* (*Il Saggiatore,* 1623):

> Philosophy is written in this grand book—I mean the universe— which stands continually open to our gaze, but it cannot be understood unless one first learns to comprehend the language and interpret the characters in which it is written. It is written in the language of mathematics, and its characters are triangles, circles, and other geometrical figures, without which it is humanly impossible to understand a single word of it; without these, one is wandering about in a dark labyrinth.

In other words, Scripture did not allow understanding; science did. The Copernican system was not a fortuitous scheme for calculating astronomical movements; it was truth.

The story of Galileo's conflict with the Church has been told many times, particularly his first trial before the Inquisition in 1616. We should note that many popular notions about this conflict are mistaken. First, Galileo was not most important as an astronomer but as a physicist. Second, the Church did not deny Galileo's astronomical observations or reject the practical utility of the Copernican system. Third, the Church did not attempt to preserve a geocentric system since that, too, would need to make room for miracles. What the Church did do was demand that Galileo acknowledge Scriptural truth. He was never excommunicated or threatened with excommunication. His trial led only to an admonishment to cease spreading his beliefs since to contradict Scripture was heresy, and his second trial (1632) arose because clearly Galileo was not remaining silent. Galileo's character is captured in a

GALILEO BEFORE THE INQUISITION.

North Wind Picture Archives

famous, but probably apocryphal, story about that second trial. Having agreed finally to renounce his beliefs, he is said to have risen from his knees and muttered as he was about to withdraw "*E pur si muove*," "and yet it moves." This could be the Earth around the Sun, the Sun around its axis, or, metaphorically, the world of science that could not be held back even by the Church.

Galileo lived much of his life after 1616 in comfortable house arrest, staying with friends and patrons and continuing his research, mostly in mechanics and hydraulics. But, being true to himself, he could not—and would rather not—maintain silence. Where Kepler thought he heard the music of the spheres, Galileo offered the muttering of man.

Evangelista Torricelli: How Space Became Empty

Evangelista Torricelli (1608–1647) rose from a working-class family to become one of the great intellects of his time, accepting the protection and guidance of an uncle who was a monk, attending a Jesuit academy, serving for years as secretary to other scholars who had university posts, and ultimately succeeding Galileo, at the older man's death, as court mathematician to the Duke of Tuscany. His lasting contributions are almost all mathematical. Indeed, while Galileo had the titles of both court mathematician and court philosopher, Torricelli held only the former. But two years after gaining that post, he proposed the experiment by which his name has become part of modern science.

The Philosopher, Aristotle, had considered the very notion of a vacuum to be a logical impossibility. It is hard for us moderns to understand that when ancient writers imagined traveling to the moon pulled by a flock of yoked geese, the fantastic element was not the travel through space but the travel to such heights. If only geese were stronger and more tractable, why couldn't men fly them to the moon?

By the Renaissance, especially with the invention of the pump, it became difficult to subscribe to Aristotle's logic. Pumps, after all, created at least temporary partial vacuums. Therefore, scientists of Torricelli's day modified Aristotle's view to say that "nature abhors a vacuum." Writing in a letter in 1644, a year after his great experiment, Torricelli himself noted this problem with authority: "Many have argued that a vacuum does not exist, others claim it exists only with difficulty in spite of the repugnance of nature; I know of no one who claims it easily exists without any resistance from nature." Torricelli's great experiment had many consequences. One was a disproof of the supposed repugnance of nature for vacuums.

It was known that pumps could draw water up only from about thirty feet down. To address this strange limitation, Torricelli took up the radical notion that air might have weight. If so, perhaps the weight of air slipping through the interstices of the earth itself pushed down on the water below and made it possible for it to rise. He set out to test this bizarre hypothesis.

Torricelli imagined a cylinder thirty feet long, sealed at one end, and filled with water. He reasoned that if one upended this cylinder open-side downward at the bottom of a large tub of water, the water in the upright cylinder would flow down and out if air had no weight; however, if it had weight, air pushing down on the surface of the water in the tub would hold up the water in the cylinder. Of course, fabricating a glass tube thirty feet long was beyond the artisans of the day. In order to use a feasible cylinder, Torricelli replaced water with the densest liquid he could find, mercury. The experiment, then, required a cylinder about three feet long, a tub, and lots of mercury.

While the materials were being made and gathered, Torricelli did some calculations. Thus, when the experiment ran, he was ready. The tub was filled with mercury. The cylinder sealed at one end was filled with mercury. A stopper was placed in the open end. The cylinder was stood upright in the tub stoppered end down and then the stopper was removed. Instantly the column of mercury descended, but only a bit. Then it stopped, leaving a gap between the top of the column of mercury and the sealed end of the cylinder. And Torricelli knew what he had done.

First, with the gap at the top of the column, Torricelli had created the first sustained partial vacuum in the written history of humanity. Second, he had proved that air had weight. These two accomplishments, however, merely settled long-standing disputes. The third was a truly remarkable discovery.

The distance to the moon had been calculated, more or less accurately, since the time of Hipparchus, who already knew it to be enormously far away. It is easy enough to calculate the distance using simple trigonometry. The practical problem is understanding how to find two points on the Earth at the same time, know their relative distance, and measure the angles from them to the Moon. Given a good model for the Earth–Moon relationship, a model provided by Copernicus and accepted by Torricelli, this was possible with a high degree of accuracy. If the weight of a column of air was what held up the water that a pump drew, then, given the relative density of water and air, Torricelli knew that the column of air pushing down on the pool surrounding the rising water could be only about five miles high, assuming it was uniformly dense from bottom to top. But the moon was two hundred forty thousand miles away. And Torricelli, like other learned people of his day, knew that, too.

In the instant that the column of mercury fell and then stopped, Torricelli realized that not only did nature not abhor a vacuum, it preferred one. Most of the distance from the Earth to its nearest neighbor was empty, absolutely empty. And the realm between the other planets and the fixed stars also, in all likelihood, was empty. In that instant, Torricelli discovered outer space.

Torricelli did not stop at that moment, however. He observed his apparatus over several days and realized that fluctuations in the height of the column of mercury reflected variations in the pressure of the air. This was, then, the first barometer in the world. In his honor, to this day, a unit used to express partial pressure and barometric pressure is the torr.

But out of this world, Torricelli is much more important, because he hung the planets in the void and broke the bridge from Earth to the heavens. What would we make of Mars then?

Library of Congress

Christian Huygens: Other Earths?

One of the greatest contributors to our ideas of Mars was Christian Huygens (1629–1695), a Dutch scientist of wide importance. The son of a wealthy, learned family, he knew many of the leading western European thinkers of his own generation and those that preceded and followed. He was welcomed to Paris first in 1655, because of his family and his intellectual reputation, and lived there for the most productive period of his life, until he fell sick in 1681 and returned to Holland. In 1666, he cofounded the French Academy of Sciences. His contributions include using Galileo's discovery of the regularity of a pendulum's motion to design and build the first pendulum clock (1658), thus improving timekeeping tremendously; the study of centrifugal force; the solution of many mathematical problems concerning curvatures; and the postulation of the first wave theory of light. How did he view Mars?

Huygens, with his interest in curvatures and light, devised new methods for grinding and polishing lenses. These improved elements enabled him to construct a telescope (1655) much superior to Galileo's. With this instrument he made some fundamental discoveries, including (1655) the rings of Saturn and the existence of Saturn's largest moon, Titan. He also discerned (1656) that the nebula (meaning "cloud" in Greek) known as Orion actu-

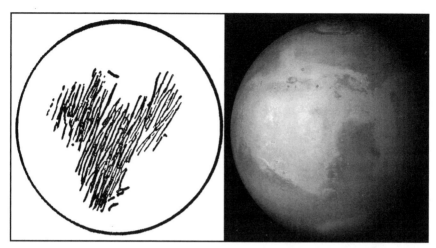

NASA, MSSS, Huygens, E. Rabkin

ally consisted of myriad stars. This vastly multiplied the number of stars one might imagine in the universe since, after all, if one apparently fuzzy star could be many, that multiplicative process might continue indefinitely. So Saturn was both like and unlike Earth—it had a moon, but it had rings— and the universe was perhaps infinite. There was room in it, then, for every imaginable oddity and for duplication of every familiarity. To Huygens, Mars offered both.

In 1659, Huygens drew what are regarded now as the first more or less accurate maps of Mars. This included a large dark spot, probably Syrtis Major, which is visible in the center of this picture taken in 1997 by the Mars Global Surveyor.

As he watched on succeeding nights, Huygens noticed not only that the dark spot moved, but that it came back to its original location in what would probably be about the duration of an Earth day. He inferred correctly that this was a demonstration of the rotation of Mars, a conclusion that both validated the idea of the Copernican system and made of Mars a close twin of the Earth.

Although Earth was not red.

In 1672, Huygens noticed the white cap on the southern pole of Mars, another first, and another feature similar to one on the Earth.

In 1698, three years after his death, Huygens' *Cosmotheros* was published. In this work, he theorizes about what a planet must provide to harbor life. He even discusses the possibility of intelligent life on other planets. Somewhere in the empty vastness of those stars upon stars, a planet like ours might harbor our cousins . . . or something much worse.

Library of Congress

Giovanni Cassini:
Very, Very Carefully

Born in Perinaldo, Italy, in 1625, Giovanni Domenico Cassini was a very careful observer. For example, using the most powerful telescopes he could order to be fabricated, by watching the shadows cast by Jupiter's satellites on the surface of that planet as they passed between the planet and the Sun, he was able to infer the period of Jupiter's rotation. He was so good an observer that in 1669, King Louis XIV of France invited Cassini to Paris where he joined the Academy Huygens had recently helped found. In 1671, on the completion of the Paris Observatory, Cassini became its director, and two years later he became Jean-Dominique Cassini, a French citizen who ultimately died in Paris in 1712 full of fame.

But Cassini was as cautious a theorist as he was careful an observer. One mustn't move too quickly. Cassini just couldn't accept Kepler's idea that the planets' orbits were ellipses, yet he could not deny the predictive power of Copernicanism. Therefore, in order to accept some version of the theory, Cassini postulated that the planets traveled on curved ovals which became known as Cassinians. He was wrong.

Since Cassini could not accept Kepler's ellipses, he could not accept Newton's Universal Law of Gravitation, which made sense of Kepler's ellipses.

Cassini Division

NASA, ESA and Erich Karkoschka (University of Arizona)

And yet Cassini discovered four more moons of Saturn to add to those Huygens had discovered twenty years earlier. He was the first to observe the "Zodiacal light," a glow made by stellar reflection off interstellar dust. Just as Brahe had rejected Copernican thought yet left observations that allowed Kepler to refine our picture of the Copernican universe, so, in 1668, Cassini compiled a table of the positions of Jupiter's satellites that the Danish astronomer Ole Rømer used in 1675 to show that the speed of light is finite and thus further refine our picture of the universe. And, most famously, Cassini observed (1675) that there was a dark band in the rings of Saturn. This band we now know is one of several, each representing a zone empty of the materials that comprise the bands, although in Cassini's day the bands were thought to be solid, like moons, rather than made of moonlets. This first-seen and most conspicuous band is called the Cassini Division.

Cassini, of course, observed Mars. In 1666, six years before Huygens noted the southern polar cap of Mars, Cassini reported the northern polar cap. Also in 1666, Cassini reported the period of Mars' rotation as 24 hours 40 minutes, which is astonishingly close to the figure we accept today, 24 hours 37 minutes 22.6 seconds. Cassini, and others, had converted Mars from a red light in the night sky into a lurid sibling of Earth.

Isaac Newton: One Big Universe

Library of Congress

How shall we regard our sibling planet? Was it, as so many had held since the Greeks, a perfect object, an ideal, something fundamentally different from our palpably corrupt Earth, or was it now simply another part of a larger system in which Earth, too, played a role, and clearly not the "starring" one? Isaac Newton supplied the answer.

Born in Lincolnshire, England, in 1642, the year of Galileo's death, Newton has often been called one of the most brilliant people who ever lived. When he died in 1727, in London, he was president of the Royal Society and the most revered thinker of his day. The importance of his fundamental accomplishments in mathematics, mechanics, optics, and astronomy were established and honored. They have not only stood the test of time but become foundational for modern science. But for understanding the role of Newton in our ideas of Mars, we need look only at his so-called Universal Law of Gravitation.

According to legend, Newton was reading beneath a tree one day when a falling apple hit his head. As Shakespeare wrote, "Ripeness is all." Newton is supposed to have realized suddenly that the same force that pulled the apple to the Earth could be one that, centripetally, pulled each planet forever around the Sun like a rock tied to the end of a whirling string. Regardless of the dubious historical truth of the story of the falling apple, Newton realized that the physical force that makes ripe apples fall, gravity, could be studied on Earth. It could operate on all objects equally (as Galileo had shown at the Tower of Pisa) and its effects could increase geometrically over time, as Galileo had also shown with his experiments using inclined planes. If this force did operate centripetally on the planets, the imaginary string holding them to a center would, as Kepler had found, map out equal areas in equal times.

What made this conclusion so stunning was not simply its immediate confirmation by existing observations but two other factors entirely. First, Newton's insight turned the Copernican system from a theory or a model into truth, just what the Inquisition had intended to suppress in its arraignment of Galileo. While Kepler did not call his own descriptions of the motions of planets "laws," following Newton, we now do. Second, Newton's law of gravitation was not a local rule applying in the sublunary world or even a grand principle applying in the superlunary world. This law applied everywhere. It was universal, which meant that what was physically, palpably true on Earth was equally true everywhere. Just because we could not reach the stars did not mean we could not understand the stars. And if the beings that Huygens was imagining might live among those stars existed, there was no reason to believe they would be angels at all, much less gods.

Newton expounded his Universal Law of Gravitation in his compendious *Philosophiae Naturalis Principia Mathematica* (*Mathematical Principles of Natural Philosophy*), commonly known as the *Principia* and published finally in 1687; however, he wrote that he did much of the original thinking that led to the *Principia* in isolation in 1665–1666 while he was at home from Cambridge to avoid the plague. At the same time and for the same reason, the poet John Dryden, who had been named a Fellow of England's Royal Society in 1662, the year of its founding, had withdrawn from London, and while rusticating he wrote a long poem called *Annus Mirabilus* (*Year of Wonders*), which was published in 1667. The wonders Dryden discusses are primarily the English commercial war against the Dutch (1665–1666) and the epochal Great Fire of London (1666). His audience, of course, also understood that the plague was an unstated wonder of that year, but Dryden doubtless left it unstated because it recalled the Biblical punishment of the Egyptians, a very unflattering comparison to make to England especially in a period of religious turmoil. (The Puritan regicide Cromwell had died in 1658 and the monarchy had been restored in 1660.) By implicit contrast with the punishing plague, the Great Fire was portrayed in terms of the mythical Phoenix: like that glorious bird, London and all England would rise again from its own ashes. The conceit moved many. In 1668, Dryden became the first person to hold the official title of Poet Laureate of England. Meanwhile Newton moved back to Cambridge.

Although Dryden did not know it, by discovering Universal Gravitation, Newton had performed yet another wonder in that year of wonders. It is pleasant then to think of Newton's fortunate bonking as occurring, if it occurred at all, in 1666, the same year in which Huygens helped found the French Academy of Science and Cassini reported Mars' northern polar cap and Mars' day to within three minutes. It is also provocative, given the theological significance of Newton's subsequent insight, to think of this legendary apple as somehow related to the forbidden fruit of the Tree of Knowledge in the Garden of Eden, and not only by the Satanic "number of the beast" (Revelations 13:18), 666, in the year 1666.

In the Bible, the serpent tempts Eve, turning her sight, like an astronomer aiming a telescope, to the fruit, which is popularly imagined to be an apple but is never actually called that in the text. "And when the woman saw that the tree was good for food, and that it was pleasant to the eyes, and a tree to be desired to make one wise, she took of the fruit thereof, and did eat, and gave also unto her husband with her; and he did eat" (Genesis 3:6). When God realizes that Adam has eaten the apple from the forbidden Tree of Knowledge, God says "Behold, the man is become as one of us" (Genesis 3:22).

To many, Newton became a hero of the intellect. To others, he helped put the institution of religion in its place: we do not need priests to explain the world to us; our own eyes and minds will do. To others, as we see in William Blake's 1795 etching called "Newton," (see color illustration at the center of the book) the scientist was a misguided hero, eyes cast downward in a sterile world, drawing his ideal but useless maps. But to Newton, who wrote pious sonnets in Latin and spent part of his life seeking to undercover the mechanism of certain Biblical prophecies, his scientific insights only showed the complete and universal perfection that everywhere pervaded God's creation. In Newton's universe, there might not be angels on Mars, but there would be no devils there either.

Mars On Their Minds

NASA/JPL/Malin Space Science Systems

Mars became a regular concern for people. It was, of course, easily visible even to amateurs and those who followed scientific controversies knew that this erstwhile god was playing key roles in the development of Enlightenment astronomy. For some astronomy was a family affair. When Cassini became blind in 1710, his son Jacques was appointed to replace him as head of the Paris Observatory. Jacques was followed by his son, César-François Cassini (also known as Cassini de Thury) who was in turn succeeded by his son, Jacques Dominique Cassini. All the Cassinis, who were appointed to their positions by the king, were devoted Royalists. Their reign in Paris ended by forced resignation in 1793, the height of the French Revolution.

While they did not concentrate particularly on Mars, a nephew of the eldest Cassini, Giacomo Filippo (later Jacques Philippe) Maraldi, did. Born in 1665 in Perinaldo, Italy, just like his uncle, he was called by the older man to assist in Paris. Maraldi arrived in 1687 and by 1702 had been elected to the French Academy of Sciences. Among his many very careful observations (he was a Cassini, after all) are records gathered at every Mars opposition from his arrival in Paris until his death there in 1729.

Opposition, which occurs when two planets line up with each other and the Sun, affords their closest mutual approach. Mars is in opposition with the Earth about every 26 months. Because the orbits of the planets are elliptical, different oppositions offer different distances of approach. During the oppositions of 1704 and 1719, Maraldi made the first well respected maps of diverse features of the planet. (In 1636 and 1638, Francisco Fontana, an amateur astronomer in Naples, Italy, drew the first and second known maps of Mars, observing merely that the planetary disk "is not uniform in color." These maps are so vaguely drawn as to be of little scientific use. Huygens had done much better in 1659.) During the opposition of 1704, Maraldi was able to confirm and slightly correct his uncle's calculations of the period of Mars' rotation, lowering the figure from 24 hours 40 minutes to the more accurate 24 hours 39 minutes. During that same opposition, Maraldi also observed what he took to be a darkening of the edges of the white polar region and suggested, again rightly, that Mars might have seasons.

In 1719, opposition brought Mars closer to Earth than it had been in about 300 years, and closer than it would be again until August, 2003. Maraldi took this occasion to study the planet well and not only produced a new map but announced the observation of a southern polar cap and correctly placed it off the circumpolar center.

But at the same time that Maraldi coolly observed the Red Planet, the public experienced hot panic. In 1719, Mars appeared so large and so red in the black night sky that it was taken by some to be a threatening comet and by others to be an ominous, inverted replay of the Star of Bethlehem. The association of Mars with some god clearly lived.

Jonathan Swift: Imaginary Travels

Literary historians debate the true beginnings of science fiction. The ancients produced candidate works, like *Icaromenippus* by Lucian of Samosata (120–180), in which Menippus acquires wings, flies to the moon, and has a revealing look down on the follies of humanity. As the title suggests, this satire transforms the much older mythic tale of Icarus, the son of the archetypal great engineer and inventor, Daedalus. Imprisoned together by King Minos in the Labyrinth of Daedelus's own construction, the father (whose name means "cunningly wrought" in Greek) made wings for himself and his son. Daedalus escaped but Icarus flew too close to the sun, melting the wax that held the feathers to his wings. He tumbled to his death near Crete into the sea that now bears his name.

There are many other early, science fiction–like works, including some by scientists, such as Kepler's *Somnium* (published posthumously in 1634) in which someone is transported to the Moon by a demon and in which the possibility of a lunar ecology is discussed. Early as they are, put together, the works of Lucian and Kepler already manifest aspects of science fiction that remain with us to this day: the notion that science may be misused, that the scientist may unleash danger even on those he most loves, that travel to new

Jonat: Swift.

Library of Congress

places may anger the gods (Apollo) or reflect the work of demons, and that from those new vantage points we can gain important new perspectives on our own selves, times, and world. If that combination marks true science fiction, one plausible candidate for the beginnings of the genre may be the third book of *Gulliver's Travels* by Jonathan Swift (1667–1745).

Gulliver's Travels (1726) is one of the most memorable and influential satires in the history of literature. Like so many other serial satires, it takes the form of a travel narrative, skewering one target after another as its viewpoint character moves from one place to another. Lemuel Gulliver, the wandering protagonist, starts out as a ship's surgeon, which is to say, since this was the early eighteenth century, that Swift's readers would not assume Gulliver to be the learned and effective healer we would expect a surgeon to be today, but still he would be more learned and certainly more observant than most of his contemporaries. The fame of the work rests most popularly, and in this order, on Gulliver's reports in Books 1, 2, and 4. Book 1, in which a shipwrecked Gulliver washes up on the shores of Lilliput, surrounds our protagonist with tiny people. Any delusions of comparative power he might have are immediately thwarted: by the time he awakens on the beach, he is already inescapably bound by numberless threads tied across him by his numerous tiny captors. One message: no man is bigger than the collectivity of society. In Book 2, Gulliver is not only powerless but small on the island of Brobdingnag where the inhabitants are as much bigger than he as he was bigger than the Lilliputians. One message: being mistaken for insignificant does not make one so. In Book 4, Gulliver travels to the land of the Houyhnms—try whinnying their name—coolly intelligent, horselike creatures who live a rational, utopian life quite obviously superior to that of the wild Yahoos—who look just like Gulliver but are speechless, filthy, and obviously brutish. One message: social good depends both on nature and intellect. But Book 3, the least read, is in some ways the most interesting for two reasons. First, it collapses the distinction between the imaginary and the real, beginning with a visit to Laputa and, four visits later, culminating with one to Japan. Second, it gives us an early critique of the scientific establishment.

Laputa is a floating island society, a huge, populated rock of 10,000 acres that rises, falls, and moves by virtue of a giant "lodestone," or magnet. Swift's more knowledgeable contemporaries would have seen immediately that Laputa is a satiric invention. The large, subdivided, supposedly utopian kingdom ruled by thoroughly distracted natural philosophers ridicules the earnest utopianism of *The New Atlantis*, the 1627 fantasy by Francis Bacon, an early and influential theorist of scientific methods, in which a 1900-year uninterrupted line of leaders from the learned house of Solamona (Solomon?) have kept society on an even keel. On Laputa (the name of which means "the whore" in Spanish, another satiric giveaway) the court astronomers literally keep the nation on an even keel as Swift makes clear in language that sounds like a report to the Royal Society.

> By means of this loadstone, the island is made to rise and fall, and move from one place to another. For, with respect to that part of the earth over which the monarch presides, the stone is endued at one of its sides with an attractive power, and at the other with a repulsive. Upon placing the magnet erect, with its attracting end towards the earth, the island descends; but when the repelling extremity points downwards, the island mounts directly upwards. When the position of the stone is oblique, the motion of the island is so too: for in this magnet, the forces always act in lines parallel to its direction.

But it must be observed, that this island cannot move beyond the extent of the dominions below, nor can it rise above the height of four miles. For which the astronomers (who have written large systems concerning the stone) assign the following reason: that the magnetic virtue does not extend beyond the distance of four miles, and that the mineral, which acts upon the stone in the bowels of the earth, and in the sea about six leagues distant from the shore, is not diffused through the whole globe, but terminated with the limits of the king's dominions; and it was easy, from the great advantage of such a superior situation, for a prince to bring under his obedience whatever country lay within the attraction of that magnet.

This loadstone is under the care of certain astronomers, who, from time to time, give it such positions as the monarch directs.

Lest this sound too ungrounded, Swift has Lemuel report that the Laputans live in large part on tribute that they haul up in baskets lowered to their subjects who are, literally, below them. And if these subjugated farmers do not care to contribute? The island is held above them, blocking the sunlight and ruining their crops. And if that is not sufficiently persuasive? Then the island is slowly lowered and the rebellion literally crushed.

Do the astronomers enabling this tyranny rebel? Quite the contrary: They spend the greatest part of their lives in observing the celestial bodies, which they do by the assistance of glasses, far excelling ours in goodness. For, although their largest telescopes do not exceed three feet, they magnify much more than those of a hundred with us, and show the stars with greater clearness. This advantage has enabled them to extend their discoveries much further than our astronomers in Europe; for they have made a catalogue of ten thousand fixed stars, whereas the largest of ours do not contain above one third part of that number. They have likewise discovered two lesser stars, or satellites, which revolve about Mars; whereof the innermost is distant from the centre of the primary planet exactly three of his diameters, and the outermost, five; the former revolves in the space of ten hours, and the latter in twenty-one and a half; so that the squares of their periodical times are very near in the same proportion with the cubes of their distance from the centre of Mars; which evidently shows them to be governed by the same law of gravitation that influences the other heavenly bodies.

The general point may be that science (nicely recalled by reference to Kepler's third law) preoccupied with the unreachable does not thereby free itself from responsibility for involvement in the world. Even a discovery as important as the moons of Mars does not validate crushing farmers. Sanitizing one's efforts by calling them science makes them no less bloody than the work they replace, that of martial enforcers.

The specific point, however, that Mars has two moons, one way or another, was probably dumb luck. Mars does have two moons, but they were discovered only in 1877, over a century and a half after *Gulliver's Travels* was published, and their size and orbits vary (although not by all that much) from those Swift notes. Perhaps he just made up the whole idea. However, in his on line *Encyclopedia of Astrobiology, Astronomy, and Spaceflight*, David Darling tells a different story:

> the idea that Mars might have two satellites goes back to Johannes Kepler and a memoir he published in 1610 in which he misconstrued an anagram devised by Galileo in order to announce secretly a new discovery to his correspondents (who also included the Jesuit Fathers at the Collegio Romano). What Galileo had actually found were features connected with the planet Saturn, which we now know to be its rings. His anagram was:
>
> s m a i s m r m i l m e p o e t a l e u m i b u n e n u g t t a u i r a s
>
> the correct solution of which was:
>
> > *Altissimum planetam tergeminum observavi.*
>
> I have observed the most distant planet [Saturn] to have a triple form. However, Kepler misconstrued the scrambled message to mean:
>
> > *Salue umbistineum geminatum Martia proles.*
>
> > Hail, twin companionship, children of Mars.
>
> and assumed, therefore, that Galileo had discovered two Martian moons. Although the true meaning of the anagram became known half a century later, Kepler's mistranslation endured and, it seems, came down to Swift.

There is, however, yet one more hypothesis for Swift's reporting the two moons of Mars. A comparison of the etching of Gulliver with one of Swift himself, especially in the eyebrows, baggy eyes, bumpy nose, fleshy chin, and slightly ironic turn of the mouth, suggests that the surgeon may have been modeled on the writer himself. But only a romantic would argue that Swift had access to the Laputans' marvelous telescopes and actually saw the two moons of Mars.

William Herschel: Stars and Mars

William Frederick Herschel (1738–1822) was not an easy man to disbelieve. Whatever others had seen, he saw more, in part because of his technical ability and devotion to observation, in part because of his bold imagination. Long before the birth of Charles Darwin (1809–1882), Herschel brought evolutionary thought to the stars and to Mars.

Herschel was born Wilhelm Friedrich Herschel in Hanover, Germany. The son of an army musician, Wilhelm expected to march to his father's beat. He joined his father's band, which was attached to the Hanoverian Guards, but after the French occupation of Hanover in 1757, young Herschel escaped to England, then ruled by George II of the House of Hanover (later renamed the House of Windsor), where he supported himself by his music, first as a music copyist, then as a teacher, performer, and even composer. Ultimately he became organist of the Octagon Chapel in the fashionable seaside resort of Bath where his house was kept by his sister Caroline who remained with him throughout his life and became his partner in his world-famous scientific research.

The love of music brought William to the study of *Harmonics, or the Philosophy of Musical Sounds* (1749) by Robert Smith (1689–1768), the Plumian

Professor of Astronomy at Trinity College, Cambridge; and then, spurred by nostalgia for nights spent star-gazing with his father, to Smith's *A Compleat System of Opticks* (1738), Herschel's first introduction to telescope making. In a refracting telescope like the one Galileo built, light is gathered and focused by a system of curved lenses made of glass. While this works well for smaller telescopes, the use of glass lenses creates a problem. As Newton discussed in his *Opticks*, light of different colors bends through glass differently. Prisms employ this principle to separate so-called white light into a rainbow. Reflecting telescopes, invented by Newton, used parabolic metal mirrors to catch and focus light, as well as glass lenses to magnify. The mirrors gather light without passing it through any medium other than air and thus create no prismatic distortion. But the reflecting mirrors before Herschel were not very effective due to the difficulty of building and polishing a parabolic mirror. Herschel changed all that.

Through experiments with many metals and alloys, Herschel so advanced the art of constructing reflecting telescopes that he soon made observations that set him apart from all others since before the dawn of history. Humanity had always known that there were seven objects that moved against the background of the stars and clearly two of these, the sun and the moon, were different from the five we now call planets. Most astronomers before Herschel concentrated on observing these seven objects. They were, after all, the only ones for which any significant detail could be viewed. But Herschel, like the astronomers of Swift's Laputa, built telescopes that "magnify much more . . . and show the stars with greater clearness." This allowed Herschel to change the very focus of astronomy.

Herschel began to study the true stars. In the course of his work, he was able to dispel the notion that there exists a luminous "fluid" in the heavens, resolving the nebulae (from the Greek word for "clouds") as collections of individual stars. He called these collections "island universes." We now call them "galaxies" (from the Greek word for "milk"). Herschel saw that the density of stars varied among these universes and so hypothesized that the stars in any given island universe attracted each other through gravity. Therefore, the denser the star field, the older the island universe. Not only did Herschel multiply the number of stars—literally—astronomically, but his foundation of the science of sidereal astronomy for the first time brought evolution into scientific thinking. The objects in the universe that we observe today reflect both the nature of Nature and the workings of time. For his many accomplishments, Herschel was knighted in 1816.

On the night of March 13, 1781, William Herschel, then still an organist by day and amateur astronomer by night, used his own handmade 20-foot reflecting telescope to make two significant observations. The less momentous was of the southern polar cap of Mars, correctly noting its size on that day, an observation that allowed him over time to track important changes in the extent of the cap. The more momentous, using his seven-foot reflecting telescope, was of a greenish star in the constellation Gemini. That night Herschel, who knew the constellation well, thought the interloper must be a comet. In subsequent months, he and other astronomers confirmed that it was no comet but the first new planet to be observed in the whole of recorded history. Herschel had not only expanded the universe but by demonstrating the vastly greater extent of the solar system, he had expanded the size of every component of the universe.

Within a year of this discovery, Herschel had been named a Fellow of the Royal Society, awarded its Copley Medal for achievement, been granted an annual pension by George III, and become Astronomer Royal. In George's honor, Herschel tried to name the new planet Georgium Sidus, the Star of George, although others suggested the planet be called Herschel. Eventually the suggestion of German astronomer Johann Elert Bode, that the planets continue to bear mythological names, won, and Herschel's Georgium Sidus became Uranus. Uranus became the most celebrated topic in astronomy and Herschel the most celebrated astronomer.

Herschel's technical ability and extraordinary knowledge never failed him. He continued to construct improved telescopes for his own use and for sale. With the support of George III, Herschel constructed what was then the world's largest telescope, an immense, forty-foot long instrument of polished wood and brass suspended in a special building we would recognize as a modern, open-roof observatory. This telescope proved too large to easily track objects, yet it was still one of the engineering wonders of the world. On the very first night of its use, Herschel discovered two previously unknown moons of Saturn.

Apparently George III invited his friend, the Archbishop of Canterbury, to see this amazing telescope in August, 1789, just as it was completed and before its fame had spread. The Archbishop, seeing this bizarre building from a distance, wondered if it were some combination of house, cannon, and sailing vessel. The King replied that it was the world's largest telescope. "Come, my lord archbishop," he is supposed to have said, "I will show you the way to heaven."

This invitation may or may not be historically accurate, but it is psychologically and culturally telling. After all, one cannot help but wonder how Herschel settled on a length of precisely forty feet for this intractable but miraculous instrument. Certainly he knew, as all Christians did, that Noah had sailed during The Flood for forty days, that the Hebrews had wandered for forty years in the desert, and that Jesus had been tempted for forty days in the wilderness. Herschel undoubtedly also knew that the Pleiades, which disappears in the latitude of Jerusalem below the horizon at about the winter solstice, reappears in the night sky forty days later. Did he also know that this visible symbol in the sky inspired the writers of the Old and New Testaments? We do not know Herschel's thoughts about this, but one way or another, King George correctly suggested that Herschel's telescope would show the way to heaven.

Herschel's career included numerous important discoveries. He demonstrated, for example, the existence of binary stars, revealed their mutual dance, and catalogued over eight hundred of them. And of course, having made the telescope, the universe, and our solar system vaster, he also explored Mars.

In 1672, Cassini had reported observing a star (Phi Aquarii) disappearing in the path of Mars a full six arc-minutes from the planet. On that basis, Cassini inferred that Mars had a thick, dense atmosphere. Herschel, however, guessed that Cassini had simply lost sight of the star in the observational glare surrounding Mars. In 1783, using his superior instrument, by observing two faint stars as Mars approached them, Herschel confirmed his hypothesis. The occultation occurred sharply at the edge of the planetary disk, demonstrating that Mars could not have the thick atmosphere Cassini described. Nonetheless, as Herschel wrote in "On the remarkable Appearances at the Polar Regions of the Planet Mars, the Inclination of its Axis, the Position of its Poles, and its spheroidical Figure; with a few Hints relating to its real Diameter and Atmosphere" (*Philosophical Transactions of the Royal Society of London* 74 [1784]), "I have often noticed occasional changes of partial bright belts . . . and also once a darkish one, in a pretty high latitude. . . . And these alterations we can hardly ascribe to any other cause than the variable disposition of clouds and vapours floating in the atmosphere of that planet." Thus Herschel inferred that Mars had an atmosphere, though, like the Earth, a thin one.

In this same report to the Royal Society, Herschel gave evidence of the seasons of Mars and, since he showed that the tilt of its axis was like that of the Earth's, Mars' seasons must echo ours, albeit at twice the length of ours since Mars' year is twice Earth's. In addition, Herschel noted that the orbit of Mars is closer to that of Earth than that of any other planet; that is, Mars bears a relation to the Sun more like ours than does any other planet. In fact, Mars was so Earth-like that Herschel speculated not only that it might support life but that its inhabitants very likely would be intelligent and "probably enjoy a situation in many respects similar to ours."

How did Herschel, who brought evolution to the stars, make this revolutionary leap of faith? Perhaps his motive in populating the sky sprang from the same well of nostalgia that held his memories of stargazing with his father the orchestra leader. Although gods and angels and even kings had been cast out of heaven, the chapel organist, brilliant scientist that he was, still heard the music of the spheres. Because he felt a great harmony between Earth and Mars, intelligent Martians entered the public imagination. Sir William Frederick Herschel was not an easy man to disbelieve.

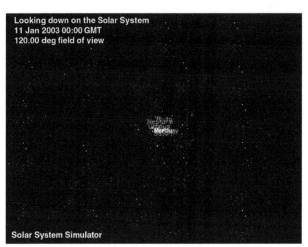

The Solar System Today

Looking down on the Solar System
11 Jan 2003 00:00 GMT
120.00 deg field of view

Nebula
Venus
Earth
Mercury

Solar System Simulator

NASA/GSFC Solar System Simulator

Thanks to William Herschel, and those on whose work he built, Mars became a neighbor of Earth, a close neighbor in a cold, vast universe. Compared to the space of stars and galaxies, our entire solar system, huge though it is compared to the world we are used to walking, seemed a tiny family huddling together about a central fire in endless, light-pricked blackness. In this awful image of the solar system, Mars, with its seasons, polar caps, and atmosphere, seemed much like Earth. By our time, with nine planets well established, the NASA Solar System Simulator cannot even label the clustering little inner planets well enough to distinguish Earth and Mars.

Asaph Hall: The Sons of Mars

Asaph Hall (1829–1907) came from a poor family in Connecticut. His father was a failed clockmaker, a profession that seems fortuitous for a son destined to demonstrate a new understanding of the clockwork of the heavens. Asaph was largely self-taught, although he did study briefly, until funds ran out, at Central College, McGrawville, N.Y., and at the University of Michigan. Despite this lack of academic credentials, he managed to convince William Cranch Bond (1789–1859), who was another self-made man of impoverished background and the director of the Harvard College Observatory, to hire him as an assistant in 1858. There Hall worked, studied, and wrote research papers, until he took up the post in 1863 of professor of mathematics at the U.S. Naval Observatory at Foggy Bottom in Washington, D.C., a position he kept until his retirement from government service in 1891. (He was later, 1896–1901, professor of astronomy at Harvard.) In 1875 Hall was given charge of the Naval Observatory's large refracting telescope. Now he could work however he liked.

According to Hall, in 1875 he had been observing the moons of Saturn when he noticed a white spot on the planet. This spot allowed him to calculate the Saturnian day. He discovered that it was a quarter of an hour different from that in the supposedly authoritative textbooks of the era. The discovery of this discrepancy, "showed [me] the necessity of consulting original papers, and made me ready to doubt the assertion one reads so often in the books, 'Mars has no moon.'" So Hall began to search.

In a solar system in which the outer planets Jupiter and Saturn have moons and in which the Earth has a moon, surely the Earth-like Mars, at an intermediate distance from the sun, would have a moon also. As Hall quickly learned, notable astronomers, including even William Herschel, had sought vainly for this new real estate. The most thorough search had been conducted in 1862 and 1864 by H. L. d'Arrest, director of the Copenhagen Observatory. D'Arrest had roughly calculated the distance at which a satellite could revolve around Mars without being flung away through its own momentum and found that this was about 70 arc-minutes as viewed from the Earth. Hall, in examining d'Arrest's work, concluded, first, that d'Arrest had confined his search to a zone bounded by that limit and, second, that d'Arrest had miscalculated the limit. According to Hall, the outer limit was about 30 arc-minutes, so he immediately began to search for moons much closer to Mars.

In August, 1877, because Edward Singleton Holden (1846–1914), his habitual assistant, was away on business visiting Henry Draper in Dobbs Ferry, New York, Hall was able to begin this quest for fame alone, just as he wished. His first observations of stars in Mars' neighborhood turned out to be just that, stars, so Hall decided to look at objects even closer to the disk of the planet. To do this, he needed to overcome the planet's glare, a feat he accomplished by "sliding the eyepiece so as to keep the planet just outside the field of view, and then turning the eyepiece in order to pass completely around the planet." The first time he tried this, he failed to discover anything, but his wife, the aptly named Angelina, pointed out that that had been a bad night to observe the heavens and urged him to persist. He did, and found something, but before he could record it properly, river fog obscured his view. Then, on August 16, he spotted it again. He called another assistant, George Anderson, to join him the next night, but, wanting the fame for himself, Hall told Anderson to "keep quiet" about what they saw. As they waited together for the reemergence of the satellite Hall had previously found, Hall observed another satellite yet closer to the planet. The two moons of Mars, the discovery that made Hall's name, had finally been seen from Earth.

However, at least two other astronomers, close associates of Hall at the Naval Observatory, sought this fame, too. Simon Newcomb (1835–1909), a senior astronomer whom Hall summoned to observe the moons, two days later claimed in print that Hall had not realized that these lights were Martian moons until Newcomb had demonstrated this mathematically. Once Hall's dated observations and calculations were published, Newcomb's case evaporated, but Newcomb maintained his centrality in this discovery his whole life, much to the anger and bitterness of his one-time colleague Hall.

At the same time that Newcomb was claiming credit in the *New York Tribune*, Holden announced in New York that he had used Draper's telescope to find a third moon of Mars. Shortly thereafter, returning to Washington, Holden claimed yet a fourth. Of course, these satellites do not exist and soon enough Holden became an object of ridicule, shamed enough in public that Hall could tolerate Holden's attempt to steal Hall's own thunder. But Hall never forgave Newcomb. It might have been impossible to stake one's claim to Martian territory in person, but money, position, and the judgment of history hung on winning that claim in the public imagination.

Accommodating in victory, Hall accepted the suggestion of Henry Madan of Eton, England, to name the moons Phobos (Fear) and Deimos (Flight or Terror), the sons of Mars whom Homer mentions in *The Iliad* as accompanying their father in battle.

Although Hall is known primarily for this one discovery in a long career, its virtue should not be minimized. Once Hall announced the existence of Phobos and Deimos, astronomers all around the world, using significantly less powerful telescopes, spotted the moons easily. Hall's success, then, had three crucial components. First, like a good, skeptical son of early New England, he had rechecked other people's work and relied on his own calculations. Second, he had let those numbers allow him to imagine something that everyone already believed just couldn't be so. And third, he developed a new technique to see if this combination of calculation and imagination just might correspond to reality. It did. Thus Hall put two more gods in the clockwork heavens.

Giovanni Schiaparelli: Gaining in Translation

U.S. Naval Observatory Library

Giovanni Virginio Schiaparelli (1835–1910) combined in himself imaginative and scientific visions of Mars. According to Schiaparelli, who was born in a poor village in the Piedmont region of Italy, near the Alps, astronomy held him from the first beautiful night when his father pointed out some of the constellations to his four-year-old son.

> Thus, as an infant, I came to know the Pleiades, the Little Wagon, the Great Wagon. . . . Also I saw the trail of a falling star; and another; and another. When I asked what they were, my father answered that this was something the Creator alone knew. Thus arose a secret and confused feeling of immense and awesome things. Already then, as later, my imagination was strongly stirred by thoughts of the vastness of space and time.

With his mother, on July 8, 1842, he watched the total eclipse of the Sun from the window of the family home, stoking his curiosity, a hunger sharpened as much as fulfilled by the rudimentary instruction he received, all the village had to offer, from Paolo Dovo, a priest who lent him books and who showed him, through a telescope in the church bell tower, his first views of the phases of Venus, the moons of Jupiter, and the rings of Saturn.

Schiaparelli, whose father was a boilermaker, graduated from the University of Turin in 1854 with a degree in architecture and hydraulic engineering. He took a job as a mathematics teacher but could not relinquish his love of astronomy. In February 1857, the government of Piedmont gave him a scholarship sufficient to allow him to study astronomy. He worked for two years at the Royal Observatory in Berlin and then for another year at the Pulkova Observatory in St. Petersburg. In 1860 he took the post of assistant astronomer at the dilapidated Brera Observatory in Milan and in 1862, at the death of Francesco Carlini, its director, Schiaparelli took Carlini's place, a position he held until his retirement in 1900. Through his extraordinary observations, he rebuilt the fame of the observatory, gathered the funding for ever more powerful equipment, and grew his institution, and his own scientific reputation, into one of the most important in the world.

Schiaparelli's discoveries were numerous. In his very first years at Brera, he demonstrated that the August falling stars followed the same orbit as a comet (1862 III, now called Swift-Tuttle), thus establishing the link between comets and what we now know to be meteor showers. Giovanni had discovered an answer previously known, according to his father, only to God. For this work, he was elected to the French Academy of Sciences and awarded the prestigious Lalande Prize. The refurbishment of Brera was underway.

His work on the comets, the asteroids, and the planets is classic, some of it remaining accepted even today, other (like the notion that Mercury's day equals its year, keeping one face always to the sun) finally being corrected as late as the 1960s. But his most famous work centered on Mars.

Schiaparelli was color blind. Perhaps for that reason, he was more sensitive to nuanced changes in light and shadow than were other observers. A handful of Mars maps had been made earlier, but they were rudimentary and disagreed with each other. Schiaparelli was able to show that some of this disagreement arose because of two phenomena. First, the clouds of Mars, which Herschel had seen, sometimes obscured features beneath them. Second, dust storms, from small ones to giants circling the whole planet, sometimes hid and sometimes revealed surface features. Schiaparelli spent night after night during the opposition of 1877, the same opposition during which Hall discovered the two small moons of Mars, observing its ever-changing surface. Finally he was able to offer the world a map.

In drawing the map, Schiaparelli followed two sets of conventions. First, he called light areas lands and dark areas seas, as had been done since the first lunar maps, even though all agreed that the seas were not liquid but

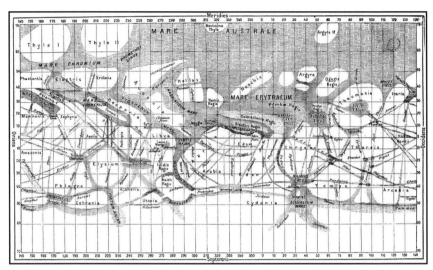

Giovanni Schiaparelli, 1877

merely areas of lower reflectance. Second, he created a nomenclature. The ancient Greek Dicaearchus (c. 326–296 B.C.E.) had drawn the first map of the Mediterranean world with a central latitude running from the Straits of Gibraltar through the Himalayas and to the assumed Eastern Ocean. Dicaearchus arrayed along this line the lands and physical features that he knew. Schiaparelli followed his lead. The mare (not really bodies of water, but so called because they were dark) took the ancient names: Mare Sirenum (Sea of Sirens), Mare Cimmerium (Sea of the Cimmerians), Mare Tyrrhenum (Tyrrhenian Sea), Mare Hadriaticum (Adriatic Sea), Syrtis Major (Gulf of Sidra), Sinus Sabaeus (Bay of Saba, today part of the Red Sea), Margaritifer Sinus (Pearl-bearing Bay, an ancient name for the rich Pearl Coast of India), Aurorae Sinus (Bay of the Dawn), and Solis Lacus (Lake of the Sun, recalling the legend according to which the Sun rises "in the baths of the ocean"). The bright lands included Ausonia (Italy, separated from Libya by the Tyrrhenian Sea), Hellas (Greece), Aeria, Arabia, Eden, Chryse, Tharsis, and Elysium.

Schiaparelli wrote that

> I do not ask that the [nomenclature] be approved by astronomers in general, nor do I request the honor of its universal acceptance. To the contrary, I am ready to accept as final whichever one is recognized by competent authority. Until then, however, grant me

the chimera of these euphonic names, whose sounds awaken in the mind so many beautiful memories.

The names stuck, in part because, in the decades following Schiaparelli's publication of his observations and map in 1878, he was the most competent authority, and in part, perhaps in larger part, because these names remade Mars as a rejuvenated classic Earthscape, a second try at a storied world that mixed shining "memories" like Hellas with a farrago of fantasies like Elysium and Eden.

The details that Schiaparelli drew, one must acknowledge, changed from time to time. While some features of Mars remained relatively fixed, others were expected to change, such as the polar caps. A handful of earlier observers had noticed fairly straight lines here and there on Mars, but the lines were few and hard to confirm. Schiaparelli, with his new, fine telescope, intense concentration, and chiaroscuro vision, saw many. Sometimes they became covered, other times revealed, but they were many and, when visible, stable. He called them, in Italian, "canali," which means "channels," an apt metaphor for lines connecting "mare" or "seas." But "canali" in Italian has a second meaning in English: "canals." From Herschel's day on, Mars had been supposed by some to be the home of intelligent beings. A simple mistranslation of Schiaparelli's great report fueled a worldwide predisposition to imagine Mars as inhabited. Where there were canals, there must be canal builders. Nostalgia and science colluded to create modern mythology. Even the most cursory glance at Schiaparelli's authoritative map demonstrated the truth of Herschel's speculation that these canal builders "enjoy a situation in many respects similar to ours."

Camille Flammarion:
Astrophile Extraordinaire

Library of Congress

Although (Nicholas) Camille Flammarion (1842–1925) was usually called an astronomer, it would be more accurate to call him an astrophile. True, he studied astronomy, and, yes, he performed some important calculations and never ceased observing; however, he did his most important astronomical work under the direction and in the service of others. His lasting contribution sprang not from his science but from his love of the stars, a romance he pursued throughout his life and which he shared with the whole world through about sixty notoriously popular books, some scientific, some fictional, most a mixture of both.

Flammarion, born poor in rural France, could not afford much education, and the education he could get did not entirely suit him. He studied theology at first, but soon turned from the heavens of the Bible to the heavens of the telescope. At sixteen, he wrote his first work (still unpublished), a five hundred page manuscript called "Universal Cosmology." His enthusiastic autodidacticism, in that same year, gained him a position at the Paris Observatory as an assistant to the autocratic LeVerrier.

Urbain Jean Joseph LeVerrier (1811–1877), a mathematician who specialized in celestial mechanics, performed calculations to explain discrepancies between Uranus's observed orbit and that predicted by the laws of Kepler and Newton. Using these calculations, and assisted by LeVerrier himself, in September, 1846, Johann Gottfried Galle (1812–1910), like Herschel before him, became the first observer of a previously unknown planet, this one less than one degree from its predicted location. The discovery of Neptune, attributed to LeVerrier's calculations rather than Galle's observational skills, made LeVerrier world-famous and encouraged him to interpret anomalous observations of the orbit of Mercury as reflecting the influence of another hypothetical planet, which he called Vulcan. Given LeVerrier's fame, this second prediction launched a raft of false detections which continued until 1915 when these anomalies were explained by Albert Einstein's General Theory of Relativity.

Flammarion, toiling under LeVerrier's direction, expressed his more creative tendencies in writing. As early as 1873, before Schiaparelli named the linear features of Mars "canali," Flammarion hypothesized that the color of Mars might be due to vegetation. In 1879 he published *L'Astronomie populaire*, a work which was widely read for decades not only in France but, in its many translations, around the world. In Flammarion's telling, astronomy was a romantic science, lifting the spirits as well as the eyes. To further encourage amateur astronomy, Flammarion founded the Astronomical Society of France in 1877, the year of Hall's discovery of the moons of Mars.

In that same year, Flammarion found and acquired the personal copy of the famous star catalog of Charles Messier (1730–1817). Flammarion deciphered Messier's handwritten notes and also used more modern instruments to identify some of Messier's objects with those Herschel had spotted and to add to Messier's catalog, ultimately issuing a revised edition. For his enthusiastic spreading of astronomy, apparently as a bolt from the blue in 1882, M. Meret, a wealthy man with no heirs, gave Flammarion a small estate at Juvisy, near Paris, to establish Flammarion's own observatory. The astrophile spent most of the rest of his life there, observing the stars as a privileged and lionized amateur and writing in factual and fictional form about his two favorite topics, life on other planets and immortality.

In 1894, *Popular Astronomy* appeared in English while in French Flammarion's *La Planète Mars* appeared, using the science of the day to argue that the canals were wrought by an advanced civilization. Ultimately, the greatest proponent of the idea of intelligent life on Mars would be an American, Percival Lowell (1855–1916), who began publishing about Mars in the following year, entering a field already well prepared by Flammarion. After Lowell's ascendancy, Flammarion, ever the amateur, became one of the American's most outspoken and noteworthy supporters.

The following report appeared in the *New York Times* of November 11,

1907. It begins with a reporter entering Flammarion's study at Juvisy. At that very moment, Flammarion is supposedly reading a letter from Lowell. Flammarion puts the letter down and delivers the following speech (to which I have added paragraph numbers). Although Flammarion is here supposedly defending Lowell, this defense is primarily an opportunity to rehearse the line of reasoning Flammarion himself had already advanced a generation earlier.

(1) My distinguished American confrère reports the existence of numerous canals in the southern hemisphere of the planet, running from the edge of the polar ice cap and joining the rest of the system in lower latitudes. These discoveries, made when the planet was in the most favorable position for observation, are just what I would have expected. It is impossible to say how these canals came into existence. They may be natural features due to the evolution of the planet, like the English Channel or the Mozambique Channel; they may be trenches dug by the inhabitants to secure the distribution of water, or they may be both, that is to say, natural formations rectified by intelligence. In any case they certainly form a most ingenious hydrographic system. It may be objected that this system does not prevent the floods which cover the plains of Mars every summer. It does not, but it regulates them, just as engineering science regulates the rising of the Nile.

(2) It is important to observe that none of the lines resembling waterways, which we call canals, seen on Mars start from land. Every one of them begins and ends either in a sea, a lake, another canal, or the intersection of several, and they cross one another at all sorts of angles. The effect is to distribute water all over the continents, and the canals are the principal, if not the only means, whereby water, and with it organic life, is spread over the surface of the planet. I think there is much to be said, although I do not care to give a positive opinion of my own, in favor of the theory that these canals and their periodical duplication are not due entirely to natural causes. The regular and geometrical appearance of the canal system helps the theory. Those who deny its possibility show a singular narrowness of mind. On the other hand, this does not justify us in supposing that life on Mars is life of the kind known to us.

(3) It is impossible to imagine what shapes are assumed by the living beings which inhabit the planet; but on the other hand, I cannot admit that the forces of nature, which are the same on Mars as with us, and are exerted under very similar conditions as to climate, atmosphere, and seasons, have been miraculously sterilized, so to speak, and rendered fruitless. On the earth the cup of life is overflowing, and the reproductive power of the human race greatly exceeds its real and durable vitality. Why should the

earth, alone among countless millions of worlds, be the only one inhabited by intelligent beings? And if we admit the force of this argument, as I think we must, why should we put Mars down as uninhabited?

(4) I believe there are denizens in Mars, and that they are superior to us, for several reasons. The first is that they could hardly be less intelligent than we are, seeing that we spend three-fourths of our resources and run heavily into debt simply to keep up armies and navies; and we cannot even agree upon a universal calendar or meridian. The second reason is that progress is an absolute, irresistible law. If the inhabitants of Mars, as we have every reason to suppose, have gone through the regular process of slow development, their present condition ought to resemble what our own will be several million years hence, inasmuch as Mars is a much older planet than the earth. Another circumstance in favor of the Martians is that they can overcome the impediment of matter far more easily than we can. The density of a cubic yard of water, or anything else, on Mars is only seven-tenths of what it is here, and a man who weights 140 pounds here would weigh only 52 on Mars. The Martian year is more than twice as long as ours, and the climatic conditions seem to be a good deal more agreeable.

(5) I dare say the Martians tried to communicate with us hundred of thousands of years ago, when mammoths were roaming around our comparatively youthful planet. The Martians may have tried again a few thousand years ago, and, never having obtained a response, they concluded that the earth was uninhabited or that its denizens did not trouble themselves about the study of the universe or the search after eternal truths. I would like to go to Mars. It must be an interesting place. No doubt the floods that cover the plains every summer would bother me at first, but one can get used to anything, and perhaps the Martians are amphibious, or know how to fly just as easily as we can walk on dry land.

(6) Of course, these are mere suppositions, but people who scout them as impossible should remember that nature is an inexhaustible mine of surprises. It was once affirmed most positively that life could not exist without oxygen, but we have since discovered creatures to which oxygen is poison. Even if Mars were without water and had a mean temperature of only 47 degrees Fahrenheit, as Prof. Lowell estimates, that would not be a good reason for calling the planet uninhabited or uninhabitable.

This speech is remarkable. It seems quite unlikely that an American reporter walked into the study only to be met immediately by such a lengthy, well organized, and uninterrupted argument. Its report in this form, then, suggests the esteem granted Flammarion. In fact, the whole article has only one brief paragraph before the speech and two after, the former situating the reporter and the debate in Juvisy and the shadow of Lowell respectively and the latter recounting the history of Flammarion's property, once known as the "Cour de France" (Court of France) because of the number of royal events staged there on official progresses between Paris and Fontainebleau. But at the time of this report, "the observatory is entirely a private institution [since] M. Flammarion's chief aim is to investigate what is not investigated elsewhere."

Paragraph (1) participates in the rhetoric of science, deploying observations as external to the observer to arrive at truth. Having tipped his beret to objectivity, Flammarion says the current findings are "just what I would have expected." Then he builds swiftly from a reasonable acknowledgment that the Martian markings may be natural to a comparison between those markings and the life-sustaining efforts of ancient as well as modern earthly engineers.

Paragraph (2) repeats this rhetorical strategy typical of Flammarion and of many other science-inspired fantasists: It begins with a statement of fact, whether it be fact or not, and then argues by stages about what is not impossible, taking at each later stage of the argument that what had been not impossible at the previous stage is now not only possible but established fact. Thus Flammarion explores the "canals" as connections among seas, lakes, and other canals, entirely ignoring the fact that Schiaparelli's nomenclature echoed the cartographic practice of Dicaearchus, an ancient Greek, and not the scientific knowledge of any modern astronomer. Nonetheless, those who do not see Mars as Flammarion does "show a singular narrowness of mind."

Paragraph (3) refers to the Martian year, atmosphere, and seasons, which were known, and "climate," which was unknown but gladly assumed by Flammarion in order to assert the habitation of Mars and disparage doubters. At this juncture, his early training and interests come out in the words "miraculous" and "fruitless": This is, as Schiaparelli's mapping of "Eden" on Mars makes plausible, a new world for creation.

Paragraph (4) uses these putative Martians to take a satiric shot at the military expenditures of modern North Atlantic nations. Flammarion's positivist faith in the "irresistible law [of] progress" matches his positivist analysis of the conditions within which Martians must have evolved. They must have great strength, for example, because Mars has low gravity. Flammarion never considers that creatures evolving under conditions of lower gravity might have commensurately less strength. Everything about Mars, in Flammarion's evolutionary view, has become grander than its equivalent on Earth. Not only are Martians our superiors, but "the climatic conditions seem to be a good deal more agreeable." At this point so much evidence has been accumulated that it is presumed we will not ask for any more to support this thoroughly subjective assertion.

Paragraph (5) leaves evidence behind entirely, projecting a romance of Martian would-be communication with fallen humans, or not yet risen humans, who are the failures of the piece. Flammarion, of course, surpasses our ancestors. He would not only reply to those futilely signaling Martians, he would want to live among them. In the instant that he imagines the "bother" he might feel from the annual summer floods, he conjures amphibious and aerial Martians who transcend floods. Flammarion does not inhabit Mars; he inhabits the realm of wish-fulfilling fiction.

Paragraph (6) again acknowledges, as any reasonable observer would, that "these are mere suppositions." However, Flammarion then reminds us of the existence of anaerobic life, which was once thought to be positively impossible, in order to ask rhetorically how we can readily reject what we believe to be not impossible. At that triumphant moment, he brings us back to Lowell, the then-reigning authority on the subject of Mars, and Flammarion rests his case.

This science fiction was reported in the *New York Times* as fact. Although Flammarion never fulfilled his wish of leaving the earth in person, in homage to his enormous stimulation of public support for their field, astronomers have placed him in the heavens nonetheless. The name of this amateur astronomer now denotes an asteroid, a lunar crater, and a crater on Mars.

Library of Congress

Percival Lowell: A Glorious Obsession

In 1910, at a Holy Cross Alumni Dinner, one John Collins Bossidy gave a toast that entered American folklore:

> *And this is to good old Boston,*
> *The home of the bean and the cod,*
> *Where the Lowells talk to the Cabots*
> *And the Cabots talk only to God.*

Most scholars have to earn fame, but Percival Lowell was born to it.

John Lowell (1743–1802), son of a well-to-do Congregational minister, was a Harvard grad (1760), a lawyer, a leader in the American Revolution, and a judge. One of his sons, Francis Cabot Lowell (1775–1817), after whom the town of Lowell, Massachusetts was named, founded the first textile mill in the United States. His grandchildren included James Russell Lowell (1819–1891), an important author and diplomat. Their cousin Augustus Lowell was the president of cotton companies and a director of banks. Augustus's wife was Katherine Bigelow Lawrence, daughter of Abbott Lawrence, a textile manufacturer and the founder of Lawrence, Massachusetts. Their children included the important poet Amy Lowell (1874–1925); A(bbott) Lawrence Lowell (1856–1943), who served as an influential president of Harvard University (1909–1933); and Percival Lowell (1855–1916), the man who got the whole world debating about "men from Mars."

Percival Lowell did not have to work, but he clearly believed what his brother Lawrence preached: "Pleasure is a by-product of doing something that is worth doing. Therefore, do not seek pleasure as such. Pleasure comes of seeking something else, and comes by the way." What did Percival seek? Strange and wonderful lands.

After graduating from Harvard (1876) with distinction in mathematics, he, like many other sons of privilege, took a Grand Tour, but Percival went farther than most, including not only Europe but parts of the Middle East. After he returned to Boston, he worked for his grandfather, John Amory Lowell, both handling mill finances and managing a mill. But Percival took to business only far enough to learn management skills and to make some excellent investments that freed him for the rest of his life from the necessity of working for money. This did not, however, deflate his desire to work or his passion for distant realms, including several trips to Asia that resulted in his being appointed as an official of a special Korean diplomatic mission to the United States (which Korea ultimately hadn't the funds to support) and his writing of four popular books, including most notably *The Soul of the Far East* (1888).

Already as a young man, Percival had been captivated by astronomy. His Harvard commencement speech concerned the nebular hypothesis of the origin of the solar system and he often brought a telescope with him on his travels. According to legend, in 1893, en route back from Japan, Percival learned that Schiaparelli's eyesight was failing (a fact even Schiaparelli seems not yet to have noticed by this date) and, fascinated by the canals and still-unglimpsed inhabitants of Mars, decided to scoop up the Italian's falling mantle as the world's leading authority on the Red Planet. Whatever the truth of Lowell's beliefs about Schiaparelli, in 1894, using personal funds, Percival Lowell of Boston, Massachusetts, moved to the then remote desert town of Flagstaff, Arizona, and built his own observatory in order to pursue what would be his life's work. Within a year, he published the first of a series of widely read and very popular books about Mars.

In the Preface to *Mars* (1895), Lowell explains the scientific reasons for establishing his observatory in Arizona and proclaims his observational discipline.

This book is the result of a special study of the planet made during the last opposition, at an observatory put up for the purpose of getting as good air as practicable, at Flagstaff, Arizona. A steady atmosphere is essential to the study of planetary detail: size of instrument being a very secondary matter. A large instrument in poor air will not begin to show what a smaller one in good air will. When this is recognized, as it eventually will be, it will become the fashion to put up observatories where they may see rather than be seen.

Next to atmosphere comes systematic study. Of the extent to which this was realized at Flagstaff, I need only say that the planet was observed there from May 24, 1894, to April 3, 1895, during which time, to mention nothing else, 917 drawings and sketches were made of it. Prof. W. H. Pickering and Mr. A. E. Douglass were associated with me in the observations herein described.

Such as care to see the original data more technically and minutely treated will find them in the first volume of the Annals of this observatory.

Lowell Observatory,
November, 1895

Thus begins the single work most influential in shaping the modern imagination of Mars. Although the language is ostensibly scientific, careful readers will notice that here Lowell projects a very personal involvement. He chastises all previous astronomers for insufficient attention to the quality of the viewing air; he claims for himself a pioneering position; he establishes his observatory in a territory that was still itself pioneering (Arizona did not become a state until Valentine's Day, 1912); he transports himself to live in a desert as Mars-like as he can find; he portrays himself as a leader of others; he places himself on the highest land he can acquire; he offers skeptics the chance to play king of the hill; and he names his institution after himself. From this grand and rugged vantage, Lowell will strive vigorously for the rest of his life to convince humanity that on Mars exist—or at least existed—its older, wiser, stronger, nobler cousins. And, despite the doubts of most—but by no means all—professional astronomers, Lowell succeeded.

Over the course of the remainder of his life, in addition to publishing *Mars,* his first major argument for life on Mars, Lowell delivered countless popular lectures, including, for example, *The Solar System: Six Lectures at the Massachusetts Institute of Technology, 1902,* published in 1903. Lowell became a professor "not in residence" at M.I.T. and his public lectures around the country were not only published after the fact but published as fact in such prestigious periodicals as the *Atlantic Monthly* and the *New York Times*. He was news. The *New York Times* (October 18, 1905) reported on the front page Lowell's successful photographic expedition with custom-made camera equipment that gave indubitable "confirmation of [Lowell's] telescopic observations." His subsequent books included *Mars and Its Canals* (1906), *Mars as the Abode of Life* (1908), and *The Evolution of Worlds* (1909).

From his aerie in Arizona, he supported an evolutionary theory, the nebular hypothesis that had been the subject of his Harvard undergraduate commencement address, in which planets formed from the cooling of gases in rings around stars, a view then being discredited. He spotted canals on Venus, and then had to withdraw his reports, and then he asserted them again. (These canals have never been observed by others.) He was sure from his observations of the perturbations in the movements of Neptune and Uranus that a more distant Planet X existed, and he supported the search for it at his magnificent observatory.

Almost all of Lowell's astronomical ideas have been discarded by science, but his decision to build observatories where they would work best set a principle that has been followed ever since. His philanthropic commitment to astronomy set an idea in the public imagination. And in 1930, fourteen years after Lowell's death, Clyde W. Tombaugh (1906–1997), then an assistant at the institution Lowell founded and endowed, following the instructions of Lowell and his associates, discovered Planet X. In keeping with the tradition of mythological naming, this planet so incredibly distant from the Sun was called Pluto, after the Greek god of the dark underworld. One of its symbols combines the first two letters of the planet's name, PL, which also happen—by no coincidence—to be the initials of Percival Lowell.

Lowell, son of one of America's most illustrious families, married late in life and died, at his observatory, without human children. But his name is in the heavens, his institution still thrives, and he is a midwife of every Martian of popular imagination. He is buried in Flagstaff, Arizona, on Mars Hill.

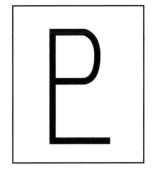

Percival Lowell: Mapping Mars and Martians

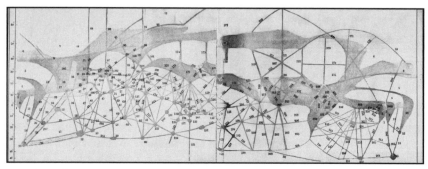

Percival Lowell, *Mars* (1895)

This map, spread over two pages in Percival Lowell's *Mars*, became a guide not only for astronomers but even more for dreamers. The classic names (referred to in Lowell's gazetteer by the reference numbers on the map), the intricate system of canals, and the authority of modern science all combined to make possible, if not a conviction in Lowell's theory of inhabited Mars then at least an eager desire for it to be true. But the map alone would have meant little without the scientific romance Lowell created of the new discovery of ancient wisdom, his pioneering certainty about a supposedly established civilization, his wilderness exploration of high culture, and his drama of youth inheriting the power of age. Lowell followed the same rhetorical strategy that Flammarion had, taking the not impossible notion of one step of reasoning as the established foundation for the next. In the Conclusion to *Mars*, Lowell recapitulates the line of argument he held in 1895 and would hold for the rest of his life, positions based equally on science and yearning—on Lowell's own astronomy, on Lowell's wish-fulfilling interpretations of cosmic and biological evolution, and even on Lowell's own ideas of what constitutes a better person and a better society, the dreams of a pioneer of privilege seeking to reunite with a glory gone, or not yet achieved, on Earth, but available in heaven. Today it may seem easy to spot Lowell's indulgences and logical blind spots, but at the dawn of the twentieth century, his argument, presented here uncut, moved millions.

CONCLUSION [LAST CHAPTER OF LOWELL'S *MARS*, 1895]

To review, now, the chain of reasoning by which we have been led to regard it probable that upon the surface of Mars we see the effects of local intelligence. We find, in the first place, that the broad physical conditions of the planet are not antagonistic to some form of life; secondly, that there is an apparent dearth of water upon the planet's surface, and therefore, if beings of sufficient intelligence inhabited it, they would have to resort to irrigation to support life; thirdly, that there turns out to be a network of markings covering the disk precisely counterparting what a system of irrigation would look like; and, lastly, that there is a set of spots placed where we should expect to find the lands thus artificially fertilized, and behaving as such constructed oases should. All this, of course, may be a set of coincidences, signifying nothing; but the probability points the other way. As to details of explanation, any we may adopt will undoubtedly be found, on closer acquaintance, to vary from the actual Martian state of things; for any Martian life must differ markedly from our own.

The fundamental fact in the matter is the dearth of water. If we keep this in mind, we shall see that many of the objections that spontaneously arise answer themselves. The supposed Herculean task of constructing such canals disappears at once; for, if the canals be dug for irrigation purposes, it is evident that what we see, and call by ellipsis the canal, is not really the canal at all, but the strip of fertilized land bordering it,—the thread of water in the midst of it, the canal itself, being far too small to be perceptible. In the case of an irrigation canal seen at a distance, it is always the strip of verdure, not the canal, that is visible, as we see in looking from afar upon irrigated country on the Earth.

We may, perhaps, in conclusion, consider for a moment how different in its details existence on Mars must be from existence on the Earth. One point out of many bearing on the subject, the simplest and most certain of all, is the effect of mere size of habitat upon the size of the inhabitant; for geometrical conditions alone are most potent factors in the problem of life. Volume and mass determine the force of gravity upon the surface of a planet, and this is more far-reaching in its effects than might at first be thought. Gravity on the surface of Mars is only a little more than one third what it is on the surface of the Earth. This would work in two ways to very different conditions of existence from those to which we are accustomed. To begin with, three times as much work, as for example, in digging a canal, could be done by the same expenditure of muscular force. If we were transported to Mars, we should be pleasingly surprised to find all our manual labor suddenly lightened threefold. But, indirectly, there might result a yet greater gain to our capabilities; for if Nature chose she could afford there to build her inhabitants on three times the scale she does on Earth without their ever finding it out except by interplanetary comparison. Let us see how.

As we all know, a large man is more unwieldy than a small one. An elephant refuses to hop like a flea; not because he considers the act undignified, but simply because he cannot bring it about. If we could, we should all jump straight across the street, instead of painfully paddling through the mud. Our inability to do so depends upon the size of the Earth, not upon what it at first seems to depend, on the size of the street.

To see this, let us consider the very simplest case, that of standing erect. To this every-day feat opposes itself the weight of the body simply, a thing of three dimensions, height, breadth, and thickness, while the ability to accomplish it resides in the cross-section of the muscles of the knee, a thing of only two dimensions, breadth and thickness. Consequently, a person half as large again as another has about twice the supporting capacity of that other, but about three times as much to support. Standing therefore tires him out more quickly. If his size were to go on increasing, he would at last reach a stature at which he would no longer be able to stand at all, but would have to lie down. You shall see the same effect in quite inanimate objects. Take two cylinders of paraffine [*sic*] wax, one made into an ordinary candle, the other into a gigantic facsimile of one, and then stand both upon their bases. To the small one nothing happens. The big one, however, begins to settle, the base actually made viscous by the pressure of the weight above.

Now apply this principle to a possible inhabitant of Mars, and suppose him to be constructed three times as large as a human being in every dimension. If he were on Earth, he would weigh twenty-seven times as much, but on the surface of Mars, since gravity there is only about one third of what it is here, he would weigh but nine times as much. The cross-section of his muscles would be nine times as great. Therefore the ratio of his supporting power to the weight he must support would be the same as ours. Consequently, he would be able to stand with as little fatigue as we. Now consider the work he might be able to do. His muscles, having length, breadth, and thickness, would all be twenty-seven times as effective as ours. He would prove twenty-seven times as strong as we, and could accomplish twenty-seven times as much. But he would further work upon what required, owing to decreased gravity, but one third the effort to overcome. His effective force, therefore, would be eighty-one times as great as man's, whether in digging canals or in other bodily occupation. As gravity on the surface of Mars is really a little more than one third that at the surface of the Earth, the true ratio is not eighty-one, but about fifty; that is, a Martian would be, physically, fiftyfold more efficient than man. As the reader will observe, there is nothing problematical about this deduction whatever. It expresses an abstract ratio of physical capabilities which must exist between the two planets, quite irrespective of whether there be denizens on either, or how other conditions may further affect their forms. As the reader must also note, the deduction refers to the possibility, not to the probability, of such giants; the calculation being introduced simply to show how different from us any Martians may be, not how different they are.

It must also be remembered that the question of their size has nothing to do with the question of their existence. The arguments for their presence are quite apart from any consideration of avoirdupois. No Herculean labors need to be accounted for; and, if they did, brain is far more potent to the task than brawn.

Something more we may deduce about the characteristics of possible Martians, dependent upon Mars itself, a result of the age of the world they would live in.

A planet may in a very real sense be said to have life of its own, of which what we call life may or may not be a subsequent detail. It is born, has its fiery youth, sobers into middle age, and just before this happens brings forth, if it be going to do so at all, the creatures on its surface which are, in a sense, its offspring. The speed with which it runs through its gamut of change prior to production depends upon its size; for the smaller the body the quicker it cools, and with it loss of heat means beginning of life for its offspring. It cools quicker because, as we saw in a previous chapter, it has relatively less inside for its outside, and it is through its outside that its inside cools. After it has thus become capable of bearing life, the Sun quickens that life and supports it for we know not how long. But its duration is measured at the most by the Sun's life. Now, inasmuch as time and space are not, as some philosophers have from their too mundane standpoint supposed, forms of our intellect, but essential attributes of the universe, the time taken by any process affects the character of the process itself, as does also the size of the body undergoing it. The changes brought about in a large planet by its cooling are not, therefore, the same as those brought about in a small one. Physically, chemically, and, to our present end, organically, the two results are quite diverse. So different, indeed, are they that unless the planet have at least a certain size it will never produce what we call life, meaning our particular chain of changes or closely allied forms of it, at all. As we saw in the case of atmosphere, it will lack even the premise to such conclusion.

Whatever the particular planet's line of development, however, in its own line, it proceeds to greater and greater degrees of evolution, till the process stops, dependent, probably, upon the Sun. The point of development attained is, as regards its capabilities, measured by the planet's own age, since the one follows upon the other.

Now, in the special case of Mars, we have before us the spectacle of a world relatively well on in years, a world much older than the Earth. To so much about his age Mars bears evidence on his face. He shows unmistakable signs of being old. Advancing planetary years have left their mark legible there. His continents are all smoothed down; his oceans have all dried up. *Teres atque rotundus* ["a man polished and complete"], he is a steady-going body now. If once he had a chaotic youth, it has long since passed away. Although called after the most turbulent of the gods, he is at the present time, whatever he may have been once, one of the most peaceable of the heavenly host. His name is a sad misnomer; indeed, the ancients seem to have been singularly unfortunate in their choice of planetary cognomens. With Mars so peaceful, Jupiter so young, and Venus bashfully draped in cloud, the planets' names accord but ill with their temperaments.

Mars being thus old himself, we know that evolution on his surface must be similarly advanced. This only informs us of its condition relative to the planet's capabilities. Of its actual state our data are not definite enough to furnish much deduction. But from the fact that our own development has been comparatively a recent thing, and that a long time would be needed to bring even Mars to his present geological condition, we may judge any life he may support to be not only relatively, but really older than our own. From the little we can see, such appears to be the case. The evidence of handicraft, if such it be, points to a highly intelligent mind behind it. Irrigation, unscientifically conducted would not give us such truly wonderful mathematical fitness in the several parts to the whole as we there behold. A mind of no mean order would seem to have presided over the system we see,—a mind certainly of considerably more comprehensiveness than that which presides over the various departments of our own public works. Party politics, at all events, have had no part in them; for the system is planet wide. Quite possibly, such Martian folk are possessed of inventions of which we have not dreamed, and with them electrophones and kinetoscopes are things of a bygone past, preserved with veneration in museums as relics of the clumsy contrivances of the simple childhood of the race. Certainly what we see hints at the existence of beings who are in advance of, not behind us, in the journey of life.

Startling as the outcome of these observations may appear at first, in truth there is nothing startling about it whatever. Such possibility has been quite on the cards ever since the existence of Mars itself was recognized by the Chaldean shepherds, or whoever the still more primeval astronomers may have been. Its strangeness is a purely subjective phenomenon, arising from the instinctive reluctance of man to admit the possibility of peers. Such would be comic were it not the inevitable consequence of the constitution of the universe. To be shy of anything resembling himself is part and parcel of man's own individuality. Like the savage who fears nothing so much as a strange man, like Crusoe who grows pale at the sight of footprints not his own, the civilized thinker instinctively turns from the thought of mind other than the one he himself knows. To admit into his conception of the cosmos other finite minds as factors has in it something of the weird. Any hypothesis to explain the facts, no matter how improbable or even palpably absurd it be, is better than this. Snow-caps of solid carbonic acid gas, a planet cracked in a positively monomaniacal manner, meteors ploughing tracks across its surface with such mathematical precision that they must have been educated to the performance, and so forth and so on, in hypotheses each more astounding than its predecessor, commend themselves to man, if only by such means he may escape the admission of anything approaching his kind. Surely all this is puerile, and should as speedily as possible be outgrown. It is simply an instinct like any other, the projection of the instinct of self-preservation. We ought, therefore, to rise above it, and, where probability points to other things, boldly accept the fact provisionally, as we should the presence of oxygen, or iron, or anything else. Let us not cheat ourselves with words. Conservatism sounds finely, and covers any amount of ignorance and fear.

We must be just as careful not to run to the other extreme, and draw deductions of purely local outgrowth. To talk of Martian beings is not to mean Martian men. Just as the probabilities point to the one, so do they point away from the other. Even on this Earth man is of the nature of an accident. He is the survival of by no means the highest physical organism. He is not even a high form of mammal. Mind has been his making. For aught we can see, some lizard or batrachian might just as well have popped into his place early in the race, and been now the dominant creature of this Earth. Under different physical conditions, he would have been certain to do so. Amid the surroundings that exist on Mars, surroundings so different from our own, we may be practically sure other organisms have been evolved of which we have no cognizance. What manner of beings they may be we lack the data even to conceive.

For answers to such problems we must look to the future. That Mars seems to be inhabited is not the last, but the first word on the subject. More important than the mere fact of the existence of living beings there, is the question of what they may be like. Whether we ourselves shall live to learn this cannot, of course, be foretold. One thing, however, we can do, and that speedily: look at things from a standpoint raised above our local point of view; free our minds at least from the shackles that of necessity tether our bodies; recognize the possibility of others in the same light that we do the certainty of ourselves. That we are the sum and substance of the capabilities of the cosmos is something so preposterous as to be exquisitely comic. We pride ourselves upon being men of the world, forgetting that this is but objectionable singularity, unless we are, in some wise, men of more worlds than one. For, after all, we are but a link in a chain. Man is merely this earth's highest production up to date. That he in any sense gauges the possibilities of the universe is humorous. He does not, as we can easily foresee, even gauge those of this planet. He has been steadily bettering from an immemorial past, and will apparently continue to improve through an incalculable future. Still less does he gauge the universe about him. He merely typifies in an imperfect way what is going on elsewhere, and what, to a mathematical certainty, is in some corners of the cosmos indefinitely excelled.

If astronomy teaches anything, it teaches that man is but a detail in the evolution of the universe, and that resemblant though diverse details are inevitably to be expected in the host of orbs around him. He learns that, though he will probably never find his double anywhere, he is destined to discover any number of cousins scattered through space.

H. G. Wells:
The War of the Worlds

Library of Congress

H(erbert) G(eorge) Wells (1866–1946) was born to a family of domestic servants who struggled to rise economically to the status of small shopkeepers. Their progress was at best uneven. Apprenticed to a draper, who let him go as unsuitable, and then to a chemist, he was largely self-taught, a prodigious reader able to digest enormous quantities of information and reveal in it patterns often undiscovered by others. By passing national examinations, he was able to win a scholarship to London University. Wells served for nearly two years, to supplement his income and advance his pattern-seeking mind, as assistant to T(homas) H(enry) Huxley (1825–1895), the great and influential disciple of Charles Darwin (1809–1882) who proselytized tirelessly for Darwin's evolutionary ideas, including the concept that from the operation of ruthless competition Nature produced ever more complex organisms, ultimately producing *Homo sapiens.* Today Wells is considered perhaps the single most important originator of the entire field of science fiction. In his heyday, from about 1895 to 1925, he was considered one of the outstanding writers of the English-speaking world, period. Throughout his brilliant career, one sees in his works a profound concern with social struggle and with evolutionary patterns.

In 1888, young Wells, then primarily a journalist, serially published *The Chronic Argonauts*. It was the first work ever to employ the narrative device of a time machine. Few noticed. In 1893, Wells published his first book, *Textbook of Biology*. He also repeatedly revised *The Chronic Argonauts*. When Wells published it again in 1895 as the now classic *The Time Machine*, he became an "overnight" success.

The Time Machine of the title transports its inventor to 802,701 A.D. where he first thinks "the whole earth had become a garden." But he soon discovers that the apathetic Eloi of the surface world, who physically remind the Time Traveler of school children, are passive prey to the hideous Morlocks of the underworld, creatures who somehow keep the hidden machinery of the world working but survive by consuming their upper-world cousins whom they snatch by night. And cousins they are. The Time Traveler deduces that the Victorian separation of the managerial class, with its pleasant life in country estates and airy town houses, had through evolutionary processes physically diverged from the working-class, consigned as they were to below-stairs domestic service and horrid travel in the underground railways. The theory of evolution outside the novel supports a dramatic representation inside the novel of a great metaphor. Even if the capitalists succeed in subjugating the workers and create for themselves what appears to be a new Eden, their very success will weaken them until implacable class conflict will turn them into cannibal food, thus debasing both descendent species. For

> [w]hat, unless biological science is a mass of errors, is the cause of human intelligence and vigour? Hardship and freedom: conditions under which the active, strong, and subtle survive and the weaker go to the wall; conditions that put a premium upon the loyal alliance of capable men, upon self-restraint, patience, and decision. And the institution of the family, and the emotions that arise therein, the fierce jealousy, the tenderness for offspring, parental self-devotion, all found their justification and support in the imminent dangers of the young.

This comment, published in the year in which Lowell published *Mars*, shares a concern for the foundational importance of evolutionary thought that Lowell displayed in his focus on the importance on Mars of the planet's age, the dearth of water, and the development of its inhabitants. Three years later, in 1898, Wells entered this arena more specifically with his publication of *The War of the Worlds*, a work that has influenced public conceptions of Mars ever since. Here is how it begins:

No one would have believed in the last years of the nineteenth century that this world was being watched keenly and closely by intelligences greater than man's and yet as mortal as his own; that as men busied themselves about their various concerns they were scrutinised and studied, perhaps almost as narrowly as a man with a microscope might scrutinise the transient creatures that swarm and multiply in a drop of water. With infinite complacency men went to and fro over this globe about their little affairs, serene in their assurance of their empire over matter. It is possible that the infusoria under the microscope do the same. No one gave a thought to the older worlds of space as sources of human danger, or thought of them only to dismiss the idea of life upon them as impossible or improbable. It is curious to recall some of the mental habits of those departed days. At most terrestrial men fancied there might be other men upon Mars, perhaps inferior to themselves and ready to welcome a missionary enterprise. Yet across the gulf of space, minds that are to our minds as ours are to those of the beasts that perish, intellects vast and cool and unsympathetic, regarded this earth with envious eyes, and slowly and surely drew their plans against us. And early in the twentieth century came the great disillusionment.

The planet Mars, I scarcely need remind the reader, revolves about the sun at a mean distance of 140,000,000 miles, and the light and heat it receives from the sun is barely half of that received by this world. It must be, if the nebular hypothesis has any truth, older than our world; and long before this earth ceased to be molten, life upon its surface must have begun its course. The fact that it is scarcely one seventh of the volume of the earth must have accelerated its cooling to the temperature at which life could begin. It has air and water and all that is necessary for the support of animated existence.

Yet so vain is man, and so blinded by his vanity, that no writer, up to the very end of the nineteenth century, expressed any idea that intelligent life might have developed there far, or indeed at all, beyond its earthly level. Nor was it generally understood that since Mars is older than our earth, with scarcely a quarter of the superficial area and remoter from the sun, it necessarily follows that it is not only more distant from time's beginning but nearer its end.

Wells here fails to acknowledge the widespread interest in Lowell's own speculations, but it must be remembered that Wells's narrator is not Wells himself but a first-person observer of the fictional future invasion who, after the fact, recounts his experiences and conclusions. Still, Wells clearly uses Lowell's evolutionary thinking and adds to it a rhetorical power that is deceptively simple. Its strength appears not only in the device of using evolution to justify a social metaphor but in his euphonious language itself, a quality made most obvious by deliberate reading aloud. Shakespeare, in *King Lear*, has doomed Gloucester observe that "As flies to wanton boys, are we to the gods, / They kill us for their sport" (IV, i, 36). One hears that echoed in Wells: "across the gulf of space, minds that are to our minds as ours are to those of the beasts that perish, intellects vast and cool and unsympathetic, regarded this earth with envious eyes, and slowly and surely drew their plans against us."

The Martians' plans, it turns out, begin quite successfully. The ships are first mistaken for meteors. When their crash landings create hot craters, people stand about considering their cooling. When, to their surprise, a cap unscrews atop the spherical "meteor" and a snakelike mechanism slowly rises, the people stand transfixed like animals caught in torch lights. But the lights on the end of this mechanism are heat rays that soon vaporize most of the gawkers. Our narrator, however, escapes, and through chance, and the application of his own knowledge, observes and understands. At one point, he watches the invaders while he is trapped in a ruined house they have occupied. The Martians are not the readily visible, huge, lethal, lumbering, crablike machines of death but the usually invisible drivers of those machines. Here is the narrator's—and Wells's—culturally decisive description of the Martians.

> They were, I now saw, the most unearthly creatures it is possible to conceive. They were huge round bodies—or, rather, heads—about four feet in diameter, each body having in front of it a face. This face had no nostrils—indeed, the Martians do not seem to have had any sense of smell, but it had a pair of very large dark-coloured eyes, and just beneath this a kind of fleshy beak. In the back of this head or body—I scarcely know how to speak of it—was the single tight tympanic surface, since known to be anatomically an ear, though it must have been almost useless in our dense air.✱ In a group round the mouth were sixteen slender, almost

---

✱ESR: This is the first of many details built on an explicit consideration of the supposed physical nature of Mars as opposed to the known physical nature of Earth.

whip-like tentacles, arranged in two bunches of eight each. These bunches have since been named rather aptly, by that distinguished anatomist, Professor Howes, the *hands*. Even as I saw these Martians for the first time they seemed to be endeavouring to raise themselves on these hands, but of course, with the increased weight of terrestrial conditions, this was impossible. There is reason to suppose that on Mars they may have progressed upon them with some facility.❋

The internal anatomy, I may remark here, as dissection has since shown, was almost equally simple. The greater part of the structure was the brain, sending enormous nerves to the eyes, ear, and tactile tentacles. Besides this were the bulky lungs,ᶜ into which the mouth opened, and the heart and its vessels. The pulmonary distress caused by the denser atmosphere and greater gravitational attraction was only too evident in the convulsive movements of the outer skin.

And this was the sum of the Martian organs. Strange as it may seem to a human being, all the complex apparatus of digestion, which makes up the bulk of our bodies, did not exist in the Martians. They were heads—merely heads. Entrails they had none. They did not eat, much less digest. Instead, they took the fresh, living blood of other creatures, and *injected* it into their own veins.✳ I have myself seen this being done, as I shall mention in its place. But, squeamish as I may seem, I cannot bring myself to describe what I could not endure even to continue watching. Let it suffice to say, blood obtained from a still living animal, in most cases from a human being, was run directly by means of a little pipette into the recipient canal. [*sic*]

The bare idea of this is no doubt horribly repulsive to us, but at the same time I think that we should remember how repulsive our carnivorous habits would seem to an intelligent rabbit.❋

❋ESR: The use of passive verb forms neatly conforms to the rhetoric of science ("There is reason to suppose" as opposed to "I suppose") and evades not only the narrator's possible idiosyncrasy as an observer but the question of how and by whom this supposition arose. Thus, as with Lowell, the not impossible of one moment becomes the granted of the next. A key difference, though, is that while both Lowell and Wells adopt the rhetoric of science, Lowell does so ostensibly to write science, while Wells does so to write fiction.

ᶜESR: The bulk of the lungs, one supposes, compensates for the thin Martian atmosphere.

✳ESR: In other words, the effects of evolution have produced Martians who are almost all brain, and vampires.

❋ESR: This comparison reflects the book's strong anti-colonial theme. As Martians are to Englishmen, so are Englishmen to Indians and other peoples subjugated by the British. As we sympathize with the humans in the novel, we should recognize the humanity of the Indians and sympathize with them. Wells uses science fiction to rebut the self-righteous

The physiological advantages of the practice of injection are undeniable, if one thinks of the tremendous waste of human time and energy occasioned by eating and the digestive process. Our bodies are half made up of glands and tubes and organs, occupied in turning heterogeneous food into blood. The digestive processes and their reaction upon the nervous system sap our strength and colour our minds. Men go happy or miserable as they have healthy or unhealthy livers, or sound gastric glands. But the Martians were lifted above all these organic fluctuations of mood and emotion.

Their undeniable preference for men as their source of nourishment is partly explained by the nature of the remains of the victims they had brought with them as provisions from Mars. These creatures, to judge from the shrivelled remains that have fallen into human hands, were bipeds with flimsy, silicious skeletons (almost like those of the silicious sponges) and feeble musculature, standing about six feet high and having round, erect heads, and large eyes in flinty sockets. Two or three of these seem to have been brought in each cylinder, and all were killed before earth was reached. It was just as well for them, for the mere attempt to stand upright upon our planet would have broken every bone in their bodies.

And while I am engaged in this description, I may add in this place certain further details which, although they were not all evident to us at the time, will enable the reader who is unacquainted with them to form a clearer picture of these offensive creatures.

In three other points their physiology differed strangely from ours. Their organisms did not sleep, any more than the heart of man sleeps. Since they had no extensive muscular mechanism to recuperate, that periodical extinction was unknown to them. They had little or no sense of fatigue, it would seem. On earth they could never have moved without effort, yet even to the last they kept in action. In twenty-four hours they did twenty-four hours of work, as even on earth is perhaps the case with the ants.✳

In the next place, wonderful as it seems in a sexual world, the Martians were absolutely without sex, and therefore without any

---

imperialism that was perhaps most famously crystallized in Rudyard Kipling's poem called "White Man's Burden": "Take up the White Man's burden, / Send forth the best ye breed— / Go, bind your sons to exile / To serve your captives' need" (1899).

✳ESR: The comparison of Martians to ants suggests that an unwavering devotion to work destroys one's individuality.

of the tumultuous emotions that arise from that difference among men. A young Martian, there can now be no dispute, was really born upon earth during the war, and it was found attached to its parent, partially *budded* off, just as young lilybulbs bud off, or like the young animals in the fresh-water polyp.✱

In man, in all the higher terrestrial animals, such a method of increase has disappeared; but even on this earth it was certainly the primitive method. Among the lower animals, up even to those first cousins of the vertebrated animals, the Tunicates, the two processes occur side by side, but finally the sexual method superseded its competitor altogether. On Mars, however, just the reverse has apparently been the case.ᶜ

It is worthy of remark that a certain speculative writer of quasi-scientific repute, writing long before the Martian invasion, did forecast for man a final structure not unlike the actual Martian condition. His prophecy, I remember, appeared in November or December, 1893, in a long-defunct publication, the *Pall Mall Budget*, and I recall a caricature of it in a pre-Martian periodical called *Punch*.✲ He pointed out—writing in a foolish, facetious

---

✱ESR: As is typical with Wells, this paragraph, which appears simple, repays close attention and, ultimately, challenges readers to make up their own minds about the issues under discussion. In this instance, we need to notice two Biblical references. The first allusion, picking up the last word of Wells's preceding paragraph, is to Proverbs 6:6–9: "Go to the ant, thou sluggard; consider her ways, and be wise: / Which having no guide, overseer, or ruler, / Provideth her meat in the summer, and gathereth her food in the harvest. / How long wilt thou sleep, O sluggard?" The first line clearly praises industriousness, and the second, about having no guide, will echo later in this excerpt from the novel, but the current passage already calls relentless work into question by its reference to the reward of work being "meat" which, in Wells's context, represents what we would view as cannibalism. The second allusion is to Matthew (6:27–30): "Which of you by taking thought can add one cubit unto his stature? / And why take ye thought for raiment? Consider the lilies of the field, how they grow; they toil not, neither do they spin: / And yet I say unto you, That even Solomon in all his glory was not arrayed like one of these. / Wherefore, if God so clothe the grass of the field, which to day is, and to morrow is cast into the oven, shall he not much more clothe you, O ye of little faith?" The traditional use of the ant to praise work and the lily to praise faith in Wells's hands serves to undercut even Solomonic thinking if it is untempered by a spirituality that grows from being in touch with one's physical reality, a connection that for Wells potentiates sympathy with others. The conclusion, that Martians—or those who behave like them—are like polyps not only places them ironically lower on Earth's evolutionary scale but reminds us of the eight-tentacled hands of the Martians. The Martian "hands," a term for workers ("how many hands work your farm?"), on Earth at least are ineffectual.

ᶜESR: Wells here offers a sexual competition between Martians and Earthlings, one between asexual and sexual reproduction. Variations on this theme of interplanetary sexual competition riddle the popular imagination of Martians.

✲ESR: This reference is to "The Man of the Year Million," an article written by Wells himself and published in the *Pall Mall Gazette*. Knowing readers of the novel doubtless enjoyed Wells's elliptical and disingenuous self-irony. The connection with the novel's anti-colonialism and with its plea for an integration of mind and body is perfect: the earlier article had portrayed our own far descendants as fundamentally all brain.

tone—that the perfection of mechanical appliances must ultimately supersede limbs; the perfection of chemical devices, digestion; that such organs as hair, external nose, teeth, ears, and
chin were no longer essential parts of the human being, and that
the tendency of natural selection would lie in the direction of
their steady diminution through the coming ages. The brain alone
remained a cardinal necessity. Only one other part of the body
had a strong case for survival, and that was the hand, "teacher
and agent of the brain." While the rest of the body dwindled, the
hands would grow larger.

There is many a true word written in jest, and here in the Martians we have beyond dispute the actual accomplishment of such a
suppression of the animal side of the organism by the intelligence.
To me it is quite credible that the Martians may be descended from
beings not unlike ourselves, by a gradual development of brain and
hands (the latter giving rise to the two bunches of delicate tentacles at last) at the expense of the rest of the body. Without the body
the brain would, of course, become a mere selfish intelligence, without any of the emotional substratum of the human being.

The last salient point in which the systems of these creatures differed from ours was in what one might have thought a very triv-
ial particular. Microorganisms, which cause so much disease and
pain on earth, have either never appeared upon Mars or Martian
sanitary science eliminated them ages ago. A hundred diseases, all
the fevers and contagions of human life, consumption, cancers,
tumours and such morbidities, never enter the scheme of their
life.• And speaking of the differences between the life on Mars
and terrestrial life, I may allude here to the curious suggestions
of the red weed.

---

*ESR: This observation, granted only half a paragraph here, becomes a crucial plot point for the Martians cannot be
resisted by humans or human technology. Nonetheless, they fail, at least this time. Near the end of the novel, the narrator finally, miraculously, sees "the Martians—*dead!*—slain by the putrefactive and disease bacteria against which their
systems were unprepared; slain as the red weed was being slain; slain, after all man's devices had failed, by the humblest
things that God, in his wisdom, has put upon this earth." That is, the (naturally or artificially) sterile atmosphere of
Mars had bred creatures who, no matter how smart, simply had not the fortitude to live in a corrupt world. Martians
on Earth, like devils in Hell, are fallen angels. But Wells is not suggesting that the answer for humanity is religion. On
the next-to-last page of the novel, the narrator says that "Dim and wonderful is the vision I have conjured up in my
mind of life spreading slowly from this little seed bed of the solar system throughout the inanimate vastness of sidereal
space. But that is a remote dream. It may be, on the other hand, that the destruction of the Martians is only a reprieve.
To them, and not to us, perhaps, is the future ordained." In other words, even if you believe in God, you must believe
in the necessity of our saving ourselves, a feat accomplished by a sympathy that arises from a full integration of brain
and body.

Apparently the vegetable kingdom in Mars, instead of having green for a dominant colour, is of a vivid blood-red tint. At any rate, the seeds which the Martians (intentionally or accidentally) brought with them gave rise in all cases to red-coloured growths. Only that known popularly as the red weed, however, gained any footing in competition with terrestrial forms. The red creeper was quite a transitory growth, and few people have seen it growing. For a time, however, the red weed grew with astonishing vigour and luxuriance. It spread up the sides of the pit by the third or fourth day of our imprisonment, and its cactus-like branches formed a carmine fringe to the edges of our triangular window. And afterwards I found it broadcast throughout the country, and especially wherever there was a stream of water. *

The Martians had what appears to have been an auditory organ, a single round drum at the back of the head–body, and eyes with a visual range not very different from ours except that, according to Philips, blue and violet were as black to them. It is commonly supposed that they communicated by sounds and tentacular gesticulations; this is asserted, for instance, in the able but hastily compiled pamphlet (written evidently by someone not an eye-witness of Martian actions) to which I have already alluded, and which, so far, has been the chief source of information concerning them. Now no surviving human being saw so much of the Martians in action as I did. I take no credit to myself for an accident, but the fact is so. And I assert that I watched them closely time after time, and that I have seen four, five, and (once) six of them sluggishly performing the most elaborately complicated operations together without either sound or gesture. Their peculiar hooting invariably preceded feeding; it had no modulation, and was, I believe, in no sense a signal, but merely the expiration of air preparatory to the suctional operation. I have a certain claim to at least an elementary knowledge of psychology, and in this matter I am convinced—as firmly as I am convinced of anything—that the Martians interchanged thoughts without any physical intermediation. And I have been convinced of this in spite of strong preconceptions. Before the Martian invasion, as an occasional reader here or there may remember, I had written with some little vehemence against the telepathic theory. ⸋

---

*ESR: Coming from the pit, the red weed, a lurid, unwanted, and invasive plant, not only reflects the color of Mars but the connection between Martians and devils.

⸋ESR: Wells here has his narrator undercut that narrator's own supposed earlier writing, making Wells's narrator the object of Wells's own irony. The jibe, however, provides more than a rhetorical flourish. Martians are, to Wells, telepathic, an ability honed by evolution. Wells's take on the implications of telepathy is that it makes Martians ant-like, they go about their work "having no guide, overseer, or ruler" and also, apparently, no individual spiritual life.

The Martians wore no clothing. Their conceptions of ornament and decorum were necessarily different from ours; and not only were they evidently much less sensible of changes of temperature than we are, but changes of pressure do not seem to have affected their health at all seriously. Yet though they wore no clothing, it was in the other artificial additions to their bodily resources that their great superiority over man lay. We men, with our bicycles and road-skates, our Lilienthal soaring-machines,* our guns and sticks and so forth, are just in the beginning of the evolution that the Martians have worked out. They have become practically mere brains, wearing different bodies according to their needs just as men wear suits of clothes and take a bicycle in a hurry or an umbrella in the wet. And of their appliances, perhaps nothing is more wonderful to a man than the curious fact that what is the dominant feature of almost all human devices in mechanism is absent—the *wheel* is absent; among all the things they brought to earth there is no trace or suggestion of their use of wheels. One would have at least expected it in locomotion. And in this connection it is curious to remark that even on this earth Nature has never hit upon the wheel, or has preferred other expedients to its development. And not only did the Martians either not know of (which is incredible), or abstain from, the wheel, but in their apparatus singularly little use is made of the fixed pivot or relatively fixed pivot, with circular motions thereabout confined to one plane. Almost all the joints of the machinery present a complicated system of sliding parts moving over small but beautifully curved friction bearings. And while upon this matter of detail, it is remarkable that the long leverages of their machines are in most cases actuated by a sort of sham musculature of the disks in an elastic sheath; these disks become polarised and drawn closely and powerfully together when traversed by a current of electricity. In this way the curious parallelism to animal motions, which was so striking and disturbing to the human beholder, was attained.ᶜ Such quasi-muscles abounded in the crablike handling-machine which, on my first peeping out of the slit, I watched unpacking the cylinder. It seemed infinitely more alive than the actual Martians lying beyond it in the sunset light, panting, stirring ineffectual tentacles, and moving feebly after their vast journey across space.

---

*ESR: The 1894 glider constructed by German engineer Otto Lilienthal (1848–1896) taught important lessons on which Wilbur and Orville Wright drew in their first successful heavier-than-air craft. At the publication of Wells's novel, however, the Wrights's success was still five years in the future, while two years earlier, Lilienthal had died in a crash of his glider. Such machines, at the moment of Wells's writing, clearly signaled human hubris.

ᶜESR: In other words, the machines so mimic living creatures and the creatures so mimic mere thinking machines that the natural integration of mind and hand with body has been "disturbing[ly]" perverted.

This image became a touchstone for our imagination of Martians: Ruthless, intellectual, sexually abhorrent, machine-like big-brains who pose an overwhelming threat to all of humanity yet can be beaten if we only will deploy our best selves and have a bit of luck. Even in this horrific take on Martians, Wells offers something uplifting, throwing the challenge of possible triumph at his readers. Millions responded. In the year of its publication in England, *The War of the Worlds* was serialized in the *Boston Post*. Young Robert Goddard (1882–1945), now generally acknowledged as the father of modern rocketry, became inspired by it. He thought of Wells's novel on October 19, 1899, which he ever after called his "Anniversary Day." According to his diary (and reminiscent, perhaps of the Fall from Eden, of Newton's apple, and of George Washington's apocryphal young exploit as a mischievous lumberjack), that day Goddard climbed a cherry tree in his backyard and "imagined how wonderful it would be to make some device which had even the *possibility* of ascending to Mars . . . when I descended the tree . . . existence at last seemed very purposive." One cannot help noticing the connection between red cherries and red Mars. For some readers, the image Wells created of Martians was not merely appealing but transforming.

"Newton," 1795, etching by William Blake. Tate Gallery, London/Art Resource, NY

Image produced by F. Hasler, M. Jentoft-Nilsen, H. Pierce, K. Palaniappan, and M. Manyin. NASA Goddard Lab for Atmospheres—Data from National Oceanic and Atmospheric Administration (NOAA)

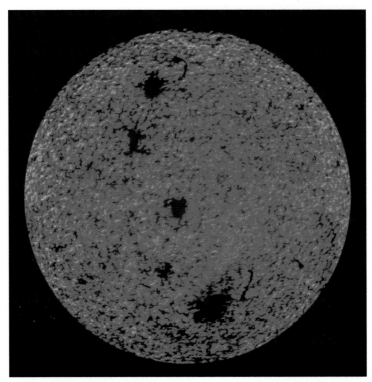

A Ball of Rust. Getty Images/PhotoDisc

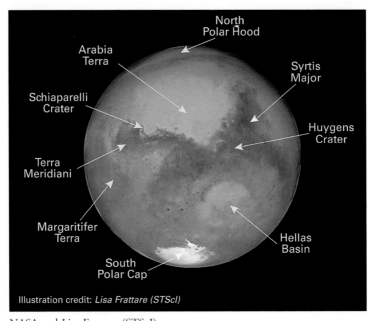

NASA and Lisa Frattare (STScI)

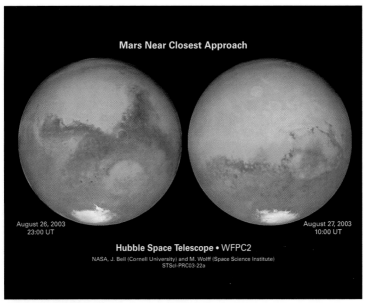

NASA, J. Bell (Cornell U.) and M. Wolff (SSI); K. Noll and A. Lubenow (STScI); M. Hubbard (Cornell U.); R. Morris (NASA/JSC); P. James (U. Toledo); S. Lee (U. Colorado); and T. Clancy, B. Whitney and G. Videen (SSI); and Y. Shkuratov (Kharkov U.)

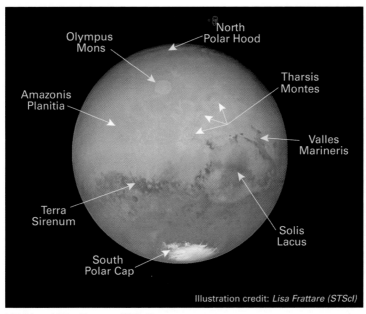

NASA and Lisa Frattare (STScI)

Movie poster. Photofest

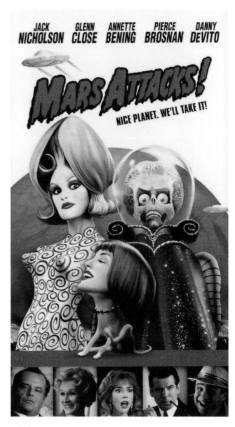

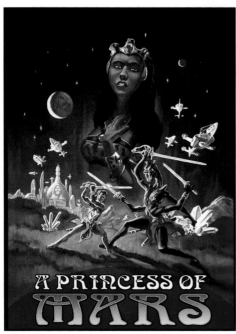

Poster. Courtesy of Jeff Doten,
http://www.jeffdoten.com

Jacket Cover from *The Martian Chronicles* by Ray Bradbury, © 1946, 1948, 1950, 1958 by Ray Bradbury; © renewed 1977 by Ray Bradbury. Used by permission of Bantam Books, a division of Random House, Inc.

Jacket Cover from *Green Mars* by Kim Stanley Robinson, © 1994. Used by permission of Bantam Books, a division of Random House, Inc.

Movie posters. Photofest

Movie poster. Photofest

Mars Rover. Jet Propulsion Laboratory

Marscape. NASA/JPL

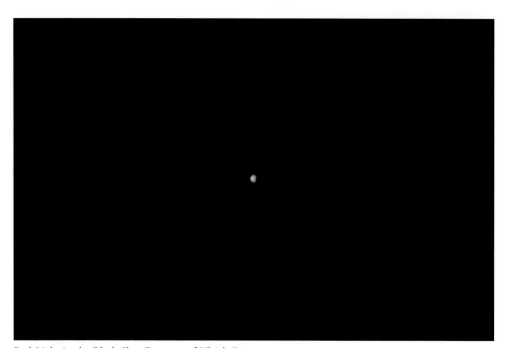

Red Light in the Black Sky. Courtesy of Ulrich Beinert

Warner Brothers Studio's Marvin the Martian. Photofest

H. G. Wells:
Another View From Mars

"The Star," a short story published by H. G. Wells in 1899, shared the quality of an elegant parable with *The Time Machine* and *The War of the Worlds* (and many of his other works from this fertile writing period such as *The Island of Dr. Moreau* [1896], *The Invisible Man* [1897], and *The Food of the Gods* [1904]). This much admired story is remarkable in many respects, one being that it has, effectively, no characters. Rather, the omniscient narrator chronicles what happens to the whole of the Earth, and to humanity, peering often into the heavens and glancing down for glimpses of an urban crowd here, an illiterate tribe there, and intermittently a person mourning, praying, or, like most of the few scientists mentioned, misunderstanding. It opens thus:

> It was on the first day of the New Year that the announcement was made, almost simultaneously from three observatories, that the motion of the planet Neptune, the outermost of all the planets that wheel about the sun, had become very erratic. Ogilvy had already called attention to a suspected retardation in its velocity

in December. Such a piece of news was scarcely calculated to interest a world the greater portion of whose inhabitants were un-
aware of the existence of the planet Neptune, nor outside the
astronomical profession did the subsequent discovery of a faint
remote speck of light in the region of the perturbed planet cause
any very great excitement. Scientific people, however, found the
intelligence remarkable enough, even before it became known
that the new body was rapidly growing larger and brighter, that
its motion was quite different from the orderly progress of the
planets, and that the deflection of Neptune and its satellite was
becoming now of an unprecedented kind.

As has so typically been the case with stories involving Mars, here, too,
we find the content of current astronomical knowledge in the forms of cultural desire. The first sentence sounds like an exact date, the quintessence of
scientific rhetoric, but the particular date is as imaginatively fraught as the
birthday of the world, which, in a way, especially divorced from any particular year, it will turn out to be. And the announcement from precisely three
observatories reminds us of the storytelling conventions of Grimm fairy tales
with their three siblings and three wishes. Scientists—people who presum-
ably define themselves by their reliance on intellect—are not too much troubled by the extraordinary observation that night, much as many people
ignored the Star of Bethlehem, which readers seeing only the story's title
might think the subject. But clearly scientists—and everyone else—should
have been troubled indeed.

In the course of the story, the "star" comes closer and closer. As it progresses, the concern and eventually panic and then destruction escalate. Mid-
way people finally recognize that this projectile into our solar system may
obliterate the Earth. Instead, it only passes nearby, although that is more
than enough. Here are the first three of the last four paragraphs of the story:

And then the clouds gathered, blotting out the vision of the sky,
the thunder and lightning wove a garment round the world; all
over the earth was such a downpour of rain as men had never before seen, and where the volcanoes flared red against the cloud
canopy there descended torrents of mud. Everywhere the waters
were pouring off the land, leaving mud-silted ruins, and the earth
littered like a storm-worn beach with all that had floated, and the
dead bodies of the men and brutes, its children. For days the
water streamed off the land, sweeping away soil and trees and
houses in the way, and piling huge dykes and scooping out Ti-

tanic gullies over the country side. Those were the days of darkness that followed the star and the heat. All through them, and for many weeks and months, the earthquakes continued.

But the star had passed, and men, hunger-driven and gathering courage only slowly, might creep back to their ruined cities, buried granaries, and sodden fields. Such few ships as had escaped the storms of that time came stunned and shattered and sounding their way cautiously through the new marks and shoals of once familiar ports. And as the storms subsided men perceived that everywhere the days were hotter than of yore, and the sun larger, and the moon, shrunk to a third of its former size, took now fourscore days between its new and new.

But of the new brotherhood that grew presently among men, of the saving of laws and books and machines, of the strange change that had come over Iceland and Greenland and the shores of Baffin's Bay, so that the sailors coming there presently found them green and gracious, and could scarce believe their eyes, this story does not tell. Nor of the movement of mankind now that the earth was hotter, northward and southward towards the poles of the earth. It concerns itself only with the coming and the passing of the Star.

Wells gives us here a scene much like that of Noah's Flood. As the Martians in *The War of the Worlds* remind us of "the scourge of God," so this "star," too, may be a scourge, an agent of ruthless cleansing to make way for a brighter new era. Indeed, this vast destruction—leading to an Earth that looks like the image of a globally warmed planet (see color illustration at the center of the book)—brings about an Edenic utopia. But Wells offers that post-apocalyptic future only as a hint. The real point of the story, after the critique of conceited science and the limits of human power, comes in the last paragraph.

The Martian astronomers—for there are astronomers on Mars, although they are very different beings from men—were naturally profoundly interested by these things. They saw them from their own standpoint of course. "Considering the mass and temperature of the missile that was flung through our solar system into the sun," one wrote, "it is astonishing what a little damage the earth, which it missed so narrowly, has sustained. All the familiar continental markings and the masses of the seas remain intact, and indeed the only difference seems to be a shrinkage of the white discoloration (supposed to be frozen water) round either pole." Which only shows how small the vastest of human catastrophes may seem, at a distance of a few million miles.

For Wells, Mars was a tool to put humanity in its place.

A World Ready to Believe

There is no doubt that one source of the success of Martian fictions like those of Wells was the world's readiness to believe. The following article appeared in the *New York Times* on November 3, 1907.

THOUGHT IT WAS A MARTIAN.

Horsefly on a Stereopticon Plate Startles Jersey Audience.

SPECIAL TO THE NEW YORK TIMES

NEW BRUNSWICK, N.J., Nov. 2.—While Prof. Robert Prentiss, astronomer of Rutgers College, was lecturing on "Mars" in the Highland Park Reformed Church last night, there was a startling demonstration of the inhabitation of the planet. The Professor had just told all about the canals which carry irrigation to the dried up planet, and was showing the canals with the aid of a stereopticon, when a monstrous thing was seen to walk upon the landscape and sit down beside a canal as though to take a drink.

As Prof. Prentiss had just told his audience that Mars was certainly inhabited, the effect was startling. The Martian had many legs, and a whole arsenal of weapons strung about him. Also he had a flying machine attachment and a horrible head, with wicked eyes that thoroughly realized H.G. Wells's conception of the inhabitants of the canal planet.

At length the Martian got up and walked up the canal several thousand miles and sat on the north pole. Prof. Prentiss, who was taken aback by the unexpected demonstration, investigated and found a horsefly on his stereopticon glass. About this time the Martian walked around to the other side of the planet and was lost to view.

Mark Wicks:
A Lowellian Utopia

Mark Wicks, *To Mars via the Moon* (1911)

The frontispiece to Mark Wicks's *To Mars via the Moon* (1911) carries three pieces of text. Just below the picture, on the left, we read *Drawn by M. Wicks*. The caption is View from the Airship, over the Canals and the City of Sirapion. There follows a text extract from the novel: "'What a splendid view we then had over the country all around us! . . . Across the country, in line after line, were the canals which we had been so anxious to see, extending as far as the eye could reach!'"

Here is the dedication of this British novel:

To
Professor Percival Lowell
A.B., LL.D.
Director of the Observatory at Flagstaff, Arizona
To whose careful and painstaking researches,

> *Extending over many years, the world owes*
> *So much of its knowledge of*
> *The planet Mars,*
> *This little book is respectfully inscribed by*
> *One who has infinite pleasure from*
> *The perusal of his works on*
> *The subject.*

As the Preface states,

> The reader will, of course, understand that whilst the astronomical information is, in all cases, scientific fact according to our present knowledge, the story itself—as well as the attempt to describe the physical and social conditions on Mars—is purely imaginative. It is not, however, merely random imagining. In a narrative such as this some matters—as, for instance, the "air-ship," and the possibility of a voyage through space—must be taken for granted; but the other ideas are mainly logical deductions from known facts and scientific data, or legitimate inferences.

But the deductions Wicks makes from Lowell are quite different from those made by Wells. Both writers pick up Lowell's idea of the advanced intelligence of the Martians. Wells construes this putative intelligence as a variety of telepathy that renders Martians machinelike and homogenized. Wicks construes this putative intelligence as a variety of telepathy, which he calls "intuition," that renders the Martians perfectly sympathetic with each other, happy and cooperative citizens of a global democracy that votes instantaneously, is served for two-year periods by an elected leader, and offers support for all according to their needs and is offered work by all according to their abilities. In short, Wells sees Martian telepathy as implying what we would call Fascism; Wicks see Martian telepathy as implying what Marx would have called Communism. Despite the effort of all concerned to assert the logic of their scientific inferences, Mars, it seems, becomes a mirror for the fears and hopes of those who appropriate it.

Wicks's Martian capital of Sirapion, with its beautiful, rectilinear plan, resembles a host of Earthly utopian cities, some of which even took shape. The sweeping, parallel diagonals converging on a governmentally crucial circle recalls the plan for Washington, D.C., completed in 1791 by Pierre-Charles L'Enfant, a French army engineer who served in the American Revolution. But whereas the name of the capital of the United States recalls that of a

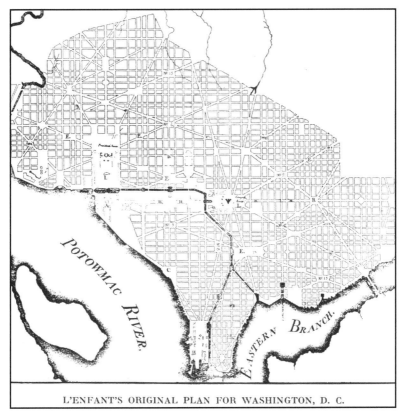

L'ENFANT'S ORIGINAL PLAN FOR WASHINGTON, D. C.

Image courtesy of the University of Wisconsin Digital Collections at http://uwdc. library.wisc.edu. Image from John Nolen's 1911 publication *Madison: A Model City*. Available online through the University of Wisconsin Digital Collections at http:// digital.library.wisc.edu/1711.dl/History.NolenMadsn

military hero, the name of the capital of Wicks's Mars recalls that of St. Serapion, a fourth century Christian who gave all his goods to others and then, repeatedly, sold himself into indentured servitude in order to earn the wherewithal to give unto others yet again. Just as Flammarion had criticized the intelligence of humans who spend fortunes on armies and navies, Wicks has the leader of Mars explain that the utopian Martians have long since outgrown war.

There are numerous features of Wicks's work that became part of our popular culture of Mars. His Martians, for instance, are seven foot nine, and parallel the "huge Martians" of the enormously influential American pulp novel, *Ralph 124 C41+* by Hugo Gernsback (published serially 1911–1912), the Hungarian immigrant who created the first science fiction magazine

(*Amazing Stories*, April, 1926) and gave the field of science fiction its name. Although *Ralph 124 C41+* is historically important for many reasons, what most concerns our discussion is that a Martian is one of the abductors of Alice, the Earthwoman who eventually becomes the bride of the title character, a handsome and strong young New York scientist. We see in Gernsback the Wellsian sexual competition and the Wicksian giant. (Gernsback may have read Wicks. *To Mars via the Moon* was available in New York, where Gernsback lived, and reviewed in the newspapers as early as April, 1911.)

The most striking feature of Wicks' appropriation of Mars is his use of it for wish fulfillment. His narrator, Wilfrid Poynders, Esq., had a son who, after three inexplicable fainting spells, dies. With his last breath he says, "I'm coming." To assuage his grief, Poynders devotes himself to getting out of this world, finally succeeding in constructing the "airship" that takes him, and two companions, to Mars. As he approaches Mars, he loses control of the vessel, but it is somehow guided inexorably to Sirapion. On the field outside the city, the Earthmen see a huge throng, almost magically assembled. They land and are greeted in English! Soon a handsome male Martian named Merna, unaccountably familiar of face, takes Poynders off for a private chat. Lo and behold! Merna is Poynders' lost son.

This encounter includes a cluster of elements that recur in "Mars Is Heaven!" (1948), a story by Ray Bradbury that became one of the key chapters of his composite novel called *The Martian Chronicles* (1950), perhaps the apotheosis of American Mars pulp fiction. The idea that Martians can control Earthmen through telepathy and guide their actions is here. (In Wicks, this extends all the way back to Earth. Merna in essence has directed his father's actions. Wells picks this up, too, in *Star Begotten* [1937]. Predictably, Wells is much darker. His Martians, a dying race, project their minds across space into human fetuses on Earth and thereby force unsuspecting women to produce essentially Martian children who aim to supplant us.) While in Wicks the dead son is truly rejoined, in Bradbury, Martian telepathy only fools the Earthmen into thinking they have re-met their lost loved ones. Come bedtime, Bradbury's trusting Earthmen are stabbed to death. This is understandable, for in post–atom bomb America, the mind is the only real weapon the Martians can use against Earth's superior, invasive technology. But for Wicks at least, the real demonstration that Mars is a mirror of desire is in the names. Merna is Poynders' son's Martian name; before his fainting death, like Wicks, he was called Mark.

Edgar Rice Burroughs:
Mars and America

If H. G. Wells, who advocated a One World socialism right here on Earth, and Mark Wicks, who fantasized about a global communism on Mars, wrote of the Red Planet out of their own British tradition, Edgar Rice Burroughs (1875–1950) seized back Percival Lowell's world of desire by imagining it as only an American could. (See color illustration at the center of the book.)

Burroughs, or ERB as he has been known to multitudes of his fans for generations, was born the son of a Chicago businessman yet lived the archetypal Horatio Alger story: rising from poverty through luck, pluck, and decency. He never seemed satisfied in any school, job, or situation, and so wound up trying his hand at many things: pursuing degrees that failed, opening small enterprises that failed, working in gold mines that failed. His most unequivocal early success was as an enlisted soldier stationed at Fort Grant, Arizona, where he worked the desert to protect the white population. He was a good soldier, perhaps in part because he developed an enormous interest in Indians and a respect for such red warrior figures as Geronimo. But when dysentery sent Burroughs to the infirmary, a heart murmur was discovered that disqualified him for an officer's commission. He left the army, tried his hand at other options, and ultimately sank so low that he had to pawn his wife's jewelry to feed the young couple. At one point, showing great faith in his talents and no acknowledgment of his failures, he cofounded a company devoted to publishing booklets on successful salesmanship, booklets he wrote himself. When that business failed, he founded a company that sold pencil sharpeners, metaphorically preparing himself for real writing. In looking through pulp magazines to make sure his pencil sharpener advertisements appeared as they should, he began to read the stories and concluded that "[i]f people are paid for writing such rot, I can write something just as rotten." He was on his way.

Burroughs' first effort became his first sale, published serially in 1911 in *All-Story Magazine* as "Under the Moons of Mars." Later reissued as *A Princess of Mars*, this was the first of Burroughs' eleven enormously popular Mars books, although they are often known by the name Burroughs' Martians gave their planet, Barsoom.

Almost immediately after the $400 acceptance of his first story, an enormous amount in 1911, Burroughs wrote "Tarzan of the Apes," which *All-Story* took for $700. Burroughs became a full-time writer.

In the course of his career, in addition to the eleven Mars books, Burroughs published seven Pellucidar (Hollow Earth) books, five Venus books, three Caspak (Forgotten Land) books, and twenty-six Tarzan books, plus about fifteen others, including one Moon book. If one counts the Tarzan books, which should not be judged by their many movie versions, as anthropological science fiction, Burroughs produced fifty-three science-fiction novels that became the central properties in a popular culture empire such as this world—Earth—had never seen before. Burroughs incorporated himself, took over his own publication, licensed the images of his characters and their plots for movies, children's paraphernalia, newspaper comic strips, comic books, and eventually radio shows and television programs. In 1919 Burroughs bought a 540-acre (0.85 square miles) ranch in the San Fernando Valley just north of Los Angeles, California, where he enjoyed being a gentleman farmer while growing his multimillion-dollar corporation. He called the ranch Tarzana, and the city that grew up around him, chartered in 1922, took the same name officially in 1927. It is still Tarzana today.

While his fortunes and influence grew, the heart murmur of his youth worsened. From 1940 on, Burroughs suffered—and deeply resented—deteriorating health. He was not a religious man, but he had his hopes. Shortly before he died, according to Erling Holtsmark, Burroughs said, "If there is a hereafter, I want to travel through space to visit other planets." We know he died in bed, of a heart attack, reading a comic book. We suspect he has gone to Mars. His ashes are buried beneath a tree in front of the office building in California that housed Edgar Rice Burroughs, Inc.

A Princess of Mars is an archetypal book that springs directly from Burroughs' experience as an individual and as an American. The opening chapter, called "On the Arizona Hills," warrants close analysis. Here is how it begins:

> I am a very old man; how old I do not know. Possibly I am a hundred, possibly more; but I cannot tell because I have never aged as other men,* nor do I remember any childhood.ᶜ So far as I can recollect I have always been a man, a man of about thirty.* I appear today as I did forty years and more ago,⁕ and yet I feel that I cannot go on living forever; that some day I shall die the real deathᶜ from which there is no resurrection.⁂ I do not know why I should fear death, I who have died twice and am still alive; but yet I have the same horror of it as you who have never died, and it is because of this terror of death, I believe, that I am so convinced of my mortality.

> And because of this conviction I have determined to write down the story of the interesting periods of my life and of my death. I cannot explain the phenomena;* I can only set down here in the words of an ordinary soldier of fortuneᶜ a chronicle of the strange events that befell me during the ten years that my dead body lay undiscovered in an Arizona cave.*

> I have never told this story, nor shall mortal man see this manuscript until after I have passed over for eternity. I know that the average human mind will not believe what it cannot grasp, and so I do not purpose being pilloried by the public, the pulpit, and the press, and held up as a colossal liar when I am but telling the simple truths which some day science will substantiate.⁕ Possibly

---

*ESR: The narrator is a man, but not a man . . .

ᶜESR: . . . and at this writing the narrator is utterly separated from knowledge of his own deepest past; that is, he is a new man who somehow has had a fresh start.

*ESR: The narrator is in his prime, at the very beginning of his wisdom and the height of his strength. He is also approximately Burroughs' age.

⁕ESR: Forty, of course, is the traditional Biblical number of death and rebirth. Is our narrator without Original Sin?

ᶜESR: "Real death" contrasts with his previous "deaths" which the narrator will recount in this book.

⁂ESR: The narrator's faith, then, despite his own rebirths, is more in himself than in a Christian god.

*ESR: Burroughs, long before he exploits the "science" of Lowell in this novel, invokes the freedom of fairy tale.

ᶜESR: "Soldier of fortune" here indicates both a mercenary, a soldier in the employ of some entity other than his government, as we will see the narrator becomes, and also indicates a soldier, which Burroughs briefly was, buffeted and boosted by fortune, that is, by fate.

*ESR: Since the narrator is clearly alive as he writes, he must have been resurrected from this cave.

⁕ESR: The future substantiation of science that promises to turn fairy tale into reality is a key claim in science fiction in general and in Mars fiction in particular.

the suggestions which I gained upon Mars, and the knowledge which I can set down in this chronicle, will aid in an earlier understanding of the mysteries of our sister planet;[*] mysteries to you, but no longer mysteries to me.

My name is John Carter;[C] I am better known as Captain Jack Carter of Virginia.[*] At the close of the Civil War I found myself possessed of several hundred thousand dollars (Confederate)[※] and a captain's commission in the cavalry arm of an army which no longer existed; the servant of a state which had vanished with the hopes of the South. Masterless, penniless, and with my only means of livelihood, fighting, gone,[C] I determined to work my way to the southwest and attempt to retrieve my fallen fortunes in a search for gold.[*]

Jack Carter and a fellow soldier are chased across the desert by Indians. The companion is fatally shot with an arrow, but Carter escapes into a cave where he is overcome by a green miasma. He falls to the ground, dead as he says at his story's commencement, and when he opens his eyes he lies somehow on the red sands of a new world. Here he discovers that his Earth-bred physique makes him more than a match for the creatures he finds, some of whom, like Dejah Thoris, the red "princess of Mars," attract him greatly. But there are green "men" who would keep Jack and Dejah Thoris apart, and cultural differences that Jack does not grasp at first in this world riven by civil war. Burroughs, in other words, creates a fairyland in which J. C. can attempt to revisit his earthly failures and pursue a sexual competition. Captain John Carter of Virginia becomes a soldier on the planet of the god of war.

Burroughs rings his own variations on many received themes. Telepathy, for example, does not cause either Fascism or Communism on Barsoom, but exists nonetheless: "Like the animals upon which the warriors were mounted the heavier draft animals wore neither bit nor bridle, but were guided by

---

[*]ESR: The notion of "sister planet" suggests Mars as a setting that offers us an alternate view of Earthly affairs. The phrase recalls Lowell's use of the term "cousins" to express the relationship between humans and Martians.

[C]ESR: The use of the initials J. C. runs throughout science fiction after Burroughs, suggesting allusively the spiritual heroism and selflessness of the hero.

[*]ESR: The more colloquial "Jack" supplants the more formal "John" and the rank of captain suggests a leader, but not a commander of multitudes. Our narrator, then, is a plain man we could know and admire. He is someone—supposing us to be adolescent males reading this story in 1911—we might wish to be. His Virgin(ia) birth makes him all the easier to identify with if one is young, Christian, or both.

[※]ESR: Like Burroughs, Carter once had means, but is now impoverished, due to losses he suffers (we are to think) through fate rather than fault of his own.

[C]ESR: Carter is a soldier of fortune with no home or opportunity to soldier.

[*]ESR: Carter follows Burroughs and the famous footsteps of the Forty-Niners and a host of other desperate, hopeful Americans who sought fame and new lives by traveling far with the aim of exploiting new lands.

telepathic means." The desirable Dejah Thoris herself offers Burroughs' pro-capitalist critique when she chides a group of the green men as "victim[s] of eons of the horrible community idea. Owning everything in common, even to your women and children, has resulted in your owning nothing in common. You hate each other as you hate all else except yourselves. Come back to the ways of our common ancestors, come back to the light of kindliness and fellowship."

Barsoom, when Carter arrives, is in the midst of an ancient, simmering civil war. While he could not succeed militarily on Earth, on Barsoom he does. This success reinforces the fundamentally conservative notions of the American Revolution, the reinstatement of the rights that all English subjects nominally held in 1776, and of the States' Rights aspirations of the Confederacy, rather than the fundamentally disruptive notions of, say, the French revolution that overturned a government to install a wholly new social experiment.

> "Was there ever such a man!" she exclaimed. "I know that Barsoom has never before seen your like. Can it be that all Earth men are as you? Alone, a stranger, hunted, threatened, persecuted, you have done in a few short months what in all the past ages of Barsoom no man has ever done: joined together the wild hordes of the sea bottoms and brought them to fight as allies of a red Martian people."

> "The answer is easy, Dejah Thoris," I replied smiling. "It was not I who did it, it was love, love for Dejah Thoris, a power that would work greater miracles than this you have seen."

> A pretty flush overspread her face and she answered,

> "You may say that now, John Carter, and I may listen, for I am free."

Here, near the book's end, we understand that Carter has become admirable because he could win a civil war through the revival of old virtues, just what he could not do on Earth. In the process, Dejah Thoris is released from the social hold of another and the swashbuckling romance of Mars turns to love. This love is truly "miraculous" because Dejah Thoris' species is egg-laying. Nonetheless, J. C. and D. T. marry, couple, and conceive. But just as mysteriously as Carter arrived on Mars, so he swoons and reanimates on Earth. The last chapter, called "At the Arizona Cave," ends with these ruminations:

The sight that met my eyes as I stepped out upon a small ledge which ran before the entrance of the cave filled me with consternation.

A new heaven and a new landscape met my gaze. The silvered mountains in the distance, the almost stationary moon hanging in the sky, the cacti-studded valley below me were not of Mars. I could scarcely believe my eyes, but the truth slowly forced itself upon me—I was looking upon Arizona from the same ledge from which ten years before I had gazed with longing upon Mars.

Burying my head in my arms I turned, broken, and sorrowful, down the trail from the cave.

Above me shone the red eye of Mars holding her awful secret, forty-eight million miles away.

Did the Martian reach the pump room? Did the vitalizing air reach the people of that distant planet in time to save them? Was my Dejah Thoris alive, or did her beautiful body lie cold in death beside the tiny golden incubator in the sunken garden of the inner courtyard of the palace of Tardos Mors, the jeddak of Helium?

For ten years I have waited and prayed for an answer to my questions. For ten years I have waited and prayed to be taken back to the world of my lost love. I would rather lie dead beside her there than live on Earth all those millions of terrible miles from her.

The old mine, which I found untouched, has made me fabulously wealthy; but what care I for wealth!

As I sit here tonight in my little study overlooking the Hudson, just twenty years have elapsed since I first opened my eyes upon Mars.

I can see her shining in the sky through the little window by my desk, and tonight she seems calling to me again as she has not called before since that long dead night, and I think I can see, across that awful abyss of space, a beautiful black-haired woman standing in the garden of a palace, and at her side is a little boy who puts his arm around her as she points into the sky toward the planet Earth, while at their feet is a huge and hideous creature [Carter's pet "watch-thing"] with a heart of gold.

> I believe that they are waiting there for me, and something tells
> me that I shall soon know.

John Carter succeeds as a Forty-Niner would wish to, but now his wishes are grander, deeper, older. (See color illustration at the center of the book.)

John Cawelti, in books like *Adventure, Mystery, and Romance* (1976), explicates the fundamental structure of the American Western story. Set in a vast, rugged, unpopulated landscape, there is always an "in" group (the townspeople, the farmers, the wagon train) and an "out" group (the outlaws, the ranchers, the Indians). The "in" group is likely to have social institutions like schools, jails, and marriage while the "out" group flaunts law, depending instead on well-honed survival skills. Into a world of ongoing conflict between these groups steps a lone hero who for some reason (typically some earlier personal loss) shares the values of the "in" group but the skills of the "out" group. He uses his skills to settle the conflict in favor of the "in" group. At this point, of course, the lone hero becomes the most dangerous character in the story. He must either hang up his guns like The Virginian (the title character of Owen Wister's 1902 novel that may indeed have influenced Burroughs) or, like The Lone Ranger (and James Fennimore Cooper's Leatherstocking in the nineteenth century) ride off. At the end of *A Princess of Mars*, Jack Carter, a very human variation of a divine archetype, has done both. He has settled the conflict in the Arizona-like landscape, conserved the "in" group's civilized values, and been made a truly peaceable member of the group through marriage; yet, he has mysteriously left Mars, too.

The last line of the novel—"I believe that they are waiting there for me, and something tells me that I shall soon know"—announces the possibility, the promise, of a sequel. On the books that followed, on Mars and elsewhere, the Chicago lad built his empire in the Golden West. Many other writers used Mars, too, of course. The Russian Alexander Bogdanov, for example, published two politically interesting utopian Mars books, *Red Star* (1908) and *Engineer Menni* (1913), both promising the triumph of Bolshevism. The Englishman C. S. Lewis published *Out of the Silent Planet* (1938), the first volume of a religious "space trilogy" that is still read today especially by those seeking to deepen their understanding of their own Christianity. But Burroughs, in impact and staying power, outdid them all. In the process, he stamped Mars indelibly as an American landscape, a territory to which one adventurer after another could travel in search of what all Americans stereotypically seek: a chance at wealth, triumph, love, and personal renewal. That was Barsoom.

Mars: The Bar from Barsoom

Mars® is a registered trademark of Mars, Incorporated and its affiliates. This trademark is used with permission. Mars, Incorporated is not associated with Praeger. Image printed with permission of Mars, Incorporated. © Mars, Inc. 2005

The Mars bar is the centerpiece of yet another story of American commercial success, this one even greater than that of Edgar Rice Burroughs, although the connection between the candy and the Red Planet is mostly fortuitous. In 1922, Frank Mars and his son, Forrest, were in a drugstore when they came up with the idea of producing and marketing a portable malted-milk ball. The branding of the result, Milky Way bars, took advantage of the interest in astronomy that writers like Burroughs fueled and simultaneously punned on the malted milk component of the confection. The product was an immediate success. As the family business grew, it relied on a number of principles, including product innovation, to outpace the competition. In 1936 they introduced yet another candy, this one, in keeping with their astronomical branding, named after themselves. The Mars bar succeeded immediately. From then on, in America and around the world, by coincidence, "Mars" also had sweetness as one of its ready associations. By 2002, Mars Incorporated had become a global (on Earth, that is) enterprise with annual revenues in excess of $14 billion.

Dead Mars?

As early as the first decade of the twentieth century, while Percival Lowell still actively proselytized for the existence of intelligent, humanoid Martians, other, more credentialed scientists argued that Mars, if it harbored life at all, could not possibly harbor "life as we know it." In 1907, Alfred Russel Wallace (1823–1913), a much honored, pioneering British biologist who is today best known for a still-disputed question about his possible priority in formulating what we generally think of as Charles Darwin's Theory of Evolution, published *Is Mars Habitable?* Wallace's clear answer? No. Wallace, using measurements of the light reflected from Mars, inferred that the planet had a surface temperature of −35°F, which meant that any water that might exist would have to be ice. Since nearly all Earthly life depends on liquid water, clearly Mars could not be habitable.

Also, "life as we know it," or at least life as it interests most of us, is aerobic, meaning that it uses free oxygen. Oxygen, however, is a chemically active element. In the presence of water—water, of course, being a necessity without which the whole debate would be over—oxygen spontaneously combines with iron, for example, to produce rust. Perhaps, many argued, Mars once held life, or the possibility of life, but any free oxygen that it once might have had now was chemically bound, as demonstrated by the color of the planet. (See color illustration at the center of the book.)

Lowell had wisely postulated red vegetation on Mars. Without photosynthesis, a process found only in plants, there would not be enough oxygen to support aerobic life on Earth. Oxygen is a by-product of photosynthesis, a process that depends on the green molecule chlorophyll, the same molecule that gives most plants (not mushrooms) their color. A red relative of chlorophyll, hemoglobin, is the active molecule central to the transport of oxygen in the blood of animals. But hemoglobin does not help produce the oxygen. Still, other worlds may have evolved other options.

For decades scientists tried to measure the atmospheric pressure of Mars. As late as the early 1950s, there was debate over a range as low as 24 millibars (18 torr) and as high as 90 millibars (68 torr). But since the average barometric pressure at sea level on Earth is about 1013 millibars (760 torr), "life as we know it" could not possibly exist, even if there were free oxygen, on Mars.

(In fact, we now know that the Martian atmosphere is even thinner than had been suspected. According to measurements taken at NASA's Viking 1 Lander site on the surface of Mars between 1976 and 1980, the pressure varied from 6.9 millibars to 9 millibars.)

By the 1930s, then, it was still possible for a scientist to maintain that there might be very simple, hardy, un-earthlike life on Mars, or even that once, long ago, there had been more complex life on Mars, but certainly in no serious way could one demonstrate scientifically that there was any chance whatsoever of humanoid life on Mars.

Of course, what scientists can demonstrate and the public can entertain are not necessarily the same thing.

Orson Welles:
The War of the Worlds Broadcast

Perhaps the single most famous radio broadcast of all time occurred on October 30, 1938. Orson Welles (1915–1985), then a quickly rising young media star, had recently founded Mercury Theatre on the Air, an anthology show. On this night before Halloween, with parents everywhere sewing costumes for their goblins, the world still struggling out of the Great Depression, and the threat of invasion by Germany and Japan felt by Americans to be very real, Welles took his players through Howard Koch's adaptation of H. G. Wells's *The War of the Worlds*. As the *New York Times* front page headline announced the next morning,

---

RADIO LISTENERS IN PANIC, TAKING WAR DRAMA AS FACT

Many Flee Homes to Escape 'Gas Raid From
Mars'—Phone Calls Swamp Police at
Broadcast of Wells Fantasy

But certainly Koch and Welles did not intend to spread panic. The show began with its habitual theme music and then Welles read a revised version of Wells's famous opening paragraphs, which make clear that all that follows is a report created well after the fictional fact. Still, Koch does adapt. Where Wells situates his narrative in London and its suburbs in the nineteenth century, Koch has Welles situate his in New York and its suburbs in the twentieth. Where Wells refers to "men," Koch, in an early feminist gesture, refers to "people." Where Wells, in a novel implicitly criticizing the colonialism of the British Empire, uses the term "empire" to refer to humanity's rule of Earth, Koch, in a radio play implicitly criticizing American racism, uses the more morally charged term "dominion." Perhaps most subtly and incisively, Koch transforms one of Wells's most famous phrases. "Yet across the gulf of space, minds that are to our minds as ours are to those of the beasts that perish . . . drew their plans against us" becomes "Yet across the gulf of space, minds that are to our minds as ours are to those of the beasts in the jungle . . . drew their plans against us." Where Wells in 1898 England had seen only conquest, Koch in 1938 America saw struggle, the outcome of the current conflict to which the work refers still very much in doubt.

Beyond this, Koch brilliantly adapted the story from the medium of the novel to that of the radio. By mimicking the interruptions of breaking news flashes, Koch reminded his listeners that they all were together in their separate homes, huddling by their radios, individually subject to large social forces. A careful listener would have easily noticed that the news flashes reported incidents that happened too fast for reality, like a reporter traveling from suburban New Jersey to a Manhattan broadcast studio in five minutes, but not everyone was predisposed to be careful. According to a poll taken the next day by the Columbia Broadcasting System, the network that carried the show, 38% of listeners had at one point or another thought the fiction real.

Could people really think this invasion was real? According to Koch and Welles, the Martians had landed at Grovers Mill, New Jersey, a tiny town only a few miles from Princeton. Hadley Cantril, a Princeton University sociologist, soon interviewed 137 people extensively and in 1940 published a book-length study called *The Invasion From Mars: A Study in the Psychology of Panic*. This significant study, among other features, divided the believers according to background, degree, and duration of belief. For our purposes, it is enough to let some of Cantril's informants, as he reported them, speak in their own words:

> My husband tried to calm me and said, "If this were really so, it would be on all stations," and he turned to one of the other stations and there was music. I retorted, "Nero fiddled while Rome burned."

My sister, her husband, my mother- and father-in-law were listening at home. I immediately called up the Maplewood police and asked if there was anything wrong. They answered, "We know as much as you do. Keep your radio turned in and follow the announcer's advice. There is no immediate danger in Maplewood." Naturally after that I was more scared than ever. I became hysterical and felt I was choking from the gas. We all kissed one another and felt we would all die. When I heard that gas was in the streets of Newark I called my brother and his wife and told them to get in their car and come right over so we could all be together.

I didn't do anything. I just kept listening. I thought, if this is the real thing you only die once—why get excited?

The lady from the next floor rushed downstairs, yelling to turn on the radio. I heard the explosion, people from Mars, end of world. I was very scared and everybody in the room was scared stiff too. There was nothing to do, for everything would be destroyed very soon. If I had had a little bottle of whiskey, I would have had a drink and said, "Let it go."

We had tuned in to listen to Orson Welles but when the flashes came I thought it was true. We called my brother who had gone out. He said he would be right down and drive away with us. When he came we were so excited. I felt, why can the children not be with us, if we are going to die. Then I called in to my husband: "Dan, why don't you get dressed? You don't want to die in your working clothes."

My mother took my word for it because after all I was a college graduate and she wasn't.

The announcer said a meteor had fallen from Mars and I was sure he thought that, but in the back of my head I had the idea that the meteor was just a camouflage. It was really an airplane like a Zeppelin that looked like a meteor and the Germans were attacking us with gas bombs. The airplane was built to look like a meteor just to fool people.

My only thought involving myself as a person in connection with it was a delight that if it spread to Stelton I would not have to pay the butcher's bill.

I looked in the icebox and saw some chicken left from Sunday dinner that I was saving for Monday night dinner. I said to my nephew, "We may as well eat this chicken—we won't be here in the morning."

The broadcast had us all worried but I knew it would at least scare ten years' life out of my mother-in-law.

I was swept along with it until something started to sound familiar. It was Orson Welles, of course! I felt awfully foolish, especially when I thought back and saw how fantastic even the little I did believe was.

On November 1, 1938, just two days after the broadcast, the *New York Times*, barely acknowledging the rights of free speech and showing little appreciation for experimentation with new forms of art, called editorially for wholesale reform of this new medium: radio.

TERROR BY RADIO

Radio ought to act promptly to prevent a repetition of the wave of panic in which it inundated the nation Sunday night by its "realistic" attempt to transfer to the air H. G. Wells's horror story of a mythical invasion by creatures from Mars. The inability of so many, tuning in late, to comprehend that they were listening to the account of an imaginary catastrophe has its ridiculous, even its pathetic, aspects. But the sobering fact remains that thousands, from one end of the country to the other, were frightened out of their senses, starting an incipient flight of hysterical refugees from the designated area, taxing the police and hospitals, confusing traffic and choking the usual means of communication. What began as "entertainment" might readily have ended in disaster.

The *Times* lay this panic at the feet of the producers and of the radio industry at large. All should have foreseen the reaction of "our people . . . just recovering from a psychosis brought on by fear of war" when exposed to provocations "inherent in the method of radio broadcasting as maintained at present in this country. It can only be cured by a deeply searching self-regulation in which every element of the radio industry should join."

The *Times* then compares radio to newspapers in terms a juvenile court judge might use of a dangerous delinquent.

> Radio is new* but it has adult responsibilities. It has not mastered itself or the material it uses. It does many things which the newspapers learned long ago not to do, such as mixing its news and its advertising. Newspapers know the two must be rigidly separated and plainly marked. In the broadcast of "The War of the Worlds" blood-curdling fiction was offered in exactly the manner that real news would have been given and interwoven with convincing actualities, such as an ordinary dance program, a definite locale and the titles of real officials.

To irresponsibly perform a masquerade of news in order to create "horror for the sake of thrill," the *Times* concludes, "underlines the need of careful self-searching in American broadcasting."

The scolding language here, an older medium addressing an upstart, reflects the often condescending attitude of many educated people toward popular (and hence pulp) science fiction, an attitude already signaled in the editorial's first sentence when it refers to Wells's honored and influential anticolonial fantasy as a "horror story." A bit over a week later, however, *The Nation* offered an important corrective viewpoint, one that puts the episode in perspective as a moment both in communication history and in global political history. That editorial concludes by saying that

> One of the dangerous results of the whole occurrence was the automatic suggestion of radio censorship, but fortunately the FCC shows no signs of taking the proposal seriously. For the rest, it provided a new insight into the power of radio. The disembodied voice has a far greater force than the printed word, as Hitler has discovered. If the Martian incident serves as even a slight inoculation against our next demagogue's appeal for a red hunt or an anti-Semitic drive it will have had its constructive effect.

---

*ESR: *The War of the Worlds* broadcast was heard in an estimated 12 million households. Census Bureau data indicate that the United States had approximately 131 million people living in 37 million households in 1940, so Welles's audience was large, even for a Sunday evening. The newspaper editorial assertion of radio's youth in part may reflect the relationships between the media. Station WWJ-AM in Detroit, Michigan, claims to be the oldest radio broadcaster of regularly scheduled programming. Its first transmission, lasting less than fifteen minutes, began at 8:15 PM on August 20, 1920 and was heard in an estimated 30 Detroit homes. It consisted of the playing of two phonograph records, an announcer asking the audience, "How do you get it?" and the playing of "Taps." Even at that formative moment, the direct connection between the radio voice and the listener rang through. The station was founded by the Scripps family, already owners of the *Detroit News* and eventually owners of one of the largest newspaper organizations in the world. Ever since that first transmission, like the relationships within most families, the relationships between newspapers and radio have been simultaneously cooperative and competitive.

Courtesy, Gay Huber, West Windsor Township

The inoculation this editorialist mentions constitutes one of the key functions of fantasy of all sorts. Having seen the enemy, the monster, the hideous possibility, it becomes less fearful, less threatening, almost tame. In time, we may even come to laugh at our former fears.

In 1988, the people of Grover's Mill, in sight of the water tower some of their older fellow citizens had blasted with shotguns fifty years earlier, erected a monument to this outstanding moment in their collective past.

Marvin the Martian:
Playing With Aliens

With *The War of the Worlds* broadcast and, more importantly, World War II behind them, Americans, for a moment, felt that the greatest war had been won and the great war god tamed. In the giddy aftermath of victory, sweet Mars extended its range from the candy stand in the movie house lobby to the theater screen itself, taking a place among the short cartoon films that were standard comic appetizers on the nation's neighborhood cinema bills of fare. In every case, such as the Woody Woodpecker cartoon called *Termites From Mars* (1952), the stylized, wacky, familiar, *American* character repels the silly-seeming invader. By far the most adorable of those would-be invaders was the Warner Brothers Studio character called Marvin the Martian. (See color illustration at the center of the book.)

Marvin the Martian first appeared in "Haredevil Hare" (1948). His push-broom helmet and sneakers clearly parody the crested helmet and sandals of the classical Mars as seen, for example, on the U.S. Capitol. Similarly, his feckless fist and hangdog eyes show him to be much less formidable than the composed and confident god. His small size suggests he is less perilous than petulant. In cartoon after cartoon, either Bugs Bunny (as in "Hasty Hare," 1952) or Daffy Duck (as in "Duck Dodgers in the $24\frac{1}{2}$ Century,"

1953) toy with Marvin and thwart his relentless desire to blow up the Earth with his "space modulator" so that he can enjoy a better view of Venus.

Had the ancient gods renewed their amours in the darkness of post-WWII American movie theaters, seeking mutual glimpses like teenagers who live across the street from each other but whose parents want them in their separate homes by 10:00 PM? One way or another, the image processor that is Hollywood made Marvin the Martian into a delightful family visitor.

Ray Bradbury:
An American Fairyland

In 1946, very soon after World War II ended in two mushroom clouds over Japan, young Ray Bradbury (1920–) published "The Million-Year Picnic," a story that would shape his creation of the single most significant American Mars novel of all time, *The Martian Chronicles*. This work is a composite novel, composed of sections about two-thirds of which were previously published, although some, like "Mars Is Heaven!" were slightly modified for their inclusion in the 1950 publication of the novel. These sections, some set on Earth, others on Mars, concern the movement of humanity from Earth to Mars. For Bradbury, though, humanity means Americans, and mostly white ones at that.

Early in these chronicles, the telepathic Martians are able to seduce and conquer the invaders, which is what, as Wells's metaphor had always implied, we are. But between the third and fourth Earth expeditions to Mars, the Martians virtually all expire, victims of chickenpox, a disease the Earth visitors had unwittingly unleashed. Like their Wellsian predecessors, Bradbury's Martians have no immunity to the microbes that trouble human youth. Conveniently for the Earthmen, this wipes the planet clean, or nearly so. In section after section, we see what it means to rehearse America's westward expansion, but this time without even the possibility of the blood crimes of black slavery and red genocide. Chickenpox on Mars gives white America a second chance. Chickenpox confers youth.

The sections of the novel are often very original but even when they are derivative, they warrant thought. "The Green Morning," for instance, is a lyric contemplation of the miraculous morning when the work of a trans-planted Johnny Appleseed figure sprouts like Jack's beanstalk and begins to produce the oxygen that will make human life on Mars less labored. In "The Off Season," a crass vandal of a colonist is euchred into accepting a vast tract of wasteland by one of the few surviving Martians, an elegant, masked, ancient figure. The colonist says, "'the old got to give way to the new. That's the law of give and take,'" meaning, of course, the old gives and the new takes, no compromise at all. He thrills at the thought of owning huge parcels of Martian land. His one goal: to erect the "first and most impor-tant" hot dog stand on the planet. In a parody of Emma Lazarus' poem of welcome inscribed on the base of the Statue of Liberty, the would-be cap-italist says, "'Earth. Send me your hungry and your starved. Something something—how does that poem go?'" The Martians, too wise and de-tached we imagine to laugh out loud, have poetic justice on their side. Earth's people destroy themselves in a global atomic war. No customers will come.

All of this has worked toward the chronicles' culminating with the million-year picnic itself, the story Bradbury wrote and published first. An Earth family, having used a hoarded private rocket to travel one-way to un-populated Mars as the final atomic war rages on Earth, takes a boat up a canal. As they travel, they hear a distant explosion, which the father admits was his remote demolition of their spacecraft. This redacted version of the culmination of the story and the novel begins in the gorgeous, ruined Mar-tian city where the children have chosen to camp.

"Why'd you blow up the rocket, Dad?"

"So we can't go back, ever. And so if any of those evil men ever come to Mars they won't know we're here."

[. . .] They stood there, King of the Hill, Top of the Heap, Ruler of All They Surveyed, Unimpeachable Monarchs and Presidents, trying to understand what it meant to own a world and how big a world really was.

Night came quickly in the thin atmosphere.

Against that cold, the father makes a bonfire of government bonds, business charts, and essays with titles like "Religious Prejudice" both to warm the family in the magical Martian night and to make a point.

> I'm burning a way of life, just like that way of life is being burned clean of Earth right now. Forgive me if I talk like a politician. I am, after all, a former state governor, and I was honest and they hated me for it. Life on Earth never settled down to doing anything very good. Science ran too far ahead of us too quickly, and the people got lost in a mechanical wilderness, like children making over pretty things, gadgets, helicopters, rockets; emphasizing the wrong items, emphasizing machines instead of how to run the machines. Wars got bigger and bigger and finally killed Earth. That's what the silent radio means. That's what we ran away from.

The father explains that the family will stay in this new world forever, but he has planned for this, seeing it as an opportunity to do better here than Earthmen did with the world they were given, but implying better than Europeans did with the continent they "discovered."

> "Now we're alone. We and a handful of others who'll land in a few days.
>
> [. . . .] Timothy looked at the last thing that Dad tossed in the fire. It was a map of the world, and it wrinkled and distorted itself hotly and went—flimpf—and was gone like a warm, black butterfly. Timothy turned away.
>
> "Now I'm going to show you the Martians," said Dad. "Come on, all of you. Here, Alice." He took her hand.
>
> Michael was crying loudly, and Dad picked him up and carried him, and they walked down through the ruins toward the canal.
>
> The canal. Where tomorrow or the next day their future wives would come up in a boat, small laughing girls now, with their father and mother.
>
> The night came down around them, and there were stars. But Timothy couldn't find Earth. It had already set. That was something to think about.

The family reaches the canal.

> "I've always wanted to see a Martian," said Michael. "Where are they, Dad? You promised."
>
> "There they are," said Dad, and he shifted Michael on his shoulder and pointed straight down.
>
> The Martians were there. Timothy began to shiver.
>
> The Martians were there—in the canal—reflected in the water. Timothy and Michael and Robert and Mom and Dad.
>
> The Martians stared back up at them for a long, long silent time from the rippling water.

Like immigrants coming to America who are shaped by the new land to which they give themselves and become Americans, Americans go to Mars and become Martians. This remaking of humanity fresh, while the corrupt "way of life is . . . burned clean," parallels the notion that American westward expansion allowed people to escape their debts, outrun their crimes, till virgin soil, inhabit a realm unspoiled by civilization, and seize opportunity. Bradbury's Mars is a fairyland where, as in the prefatory verses of Lewis Caroll's great novel, Alice in a gliding boat listens to a dream of "wonderland," where people survive idyllically in a landscape that 1940s science already knew could not support human life, where magical plants ease the very burden of breathing, where the white man's dirty work is fortuitously accomplished by purely accidental germ warfare. Bradbury sings an ironic variation on Wells's salvation of Earth through putrefactive microbes. But while Wells embraced science and technology, warning only that the body be integrated with the brain and hand lest we become Martians, Bradbury wants to throw technology, our mechanical hands, back a century or more so we can become Martians.

In his own life, Bradbury is famous for his technophobia. Despite living from age fourteen on in Los Angeles, perhaps the most auto dependent major city in the world, he had not learned to drive a car by the time, at his age of 62, he took his first airplane flight, an occurrence so remarkable that *Time* magazine (November 8, 1982) ran a picture of the white-knuckled author being strapped in by a smiling cabin attendant.

The pastoral cover illustration for *The Martian Chronicles* (see color illustration at the center of the book) suggests Mars as a nostalgic landscape of redemption. This book so took the public imagination that it was the first science-fiction novel ever to get a *New York Times Book Review* front page mention. It garnered Bradbury medals and awards. But many other science fiction writers of the period complained that Bradbury was no *science*-fiction writer at all, since he shunned science so obviously. He had, though, re-fashioned large tracts of Mars and made a stand that filled a need for those "hungry and . . . starved" for a simpler, easier time of hope and growth, for a time of personal and national youth, for a moment when the distant clouds were not atomic blasts towering over immolated cities but mists hovering above peaceful canals.

George Pal:
The War of the Worlds again

George Pal (1908–1980) was perhaps the most polished of the makers of 1950s Hollywood science fiction films. A Hungarian immigrant who fled Nazi Europe, Pal developed the idea of "puppetoons" (puppets plus cartoons), three-dimensional animations that he used in witty short movies and for which he won a special effects Academy Award in 1943. He went on to win five more Academy Awards, including one for his rendition of *The War of the Worlds* (1953; see color illustration at the center of the book).

Most of the many Hollywood science fiction films of the fifteen post-WWII years were low-budget works that played on American fears. While horror movies tended to play on personal fears, like the dangerous sexuality of the many vampire movies, science fiction tended to work with public fears. In B-budget movie after B-budget movie, we see vivid images of xenophobia (*It Came From Outer Space*, 1953, produced by Jack Arnold and scripted by Ray Bradbury); fear of technology unbridled by social responsibility (*Tarantula*, 1955, also by Arnold, in which a scientist's experimental nutrient produces a hundred-foot-wide rampaging spider); fear of a Communist fifth column (*Invasion of the Body Snatchers*, 1955, in which hive-minded pod people remake themselves to supplant and pass for our neighbors); and sometimes all three fears at once (*Them!* 1954, in which atomic bomb tests mutate ants to become six-foot terrors that nest in their homogenous implacability beneath the sewers of Los Angeles just waiting to hatch and destroy—gasp!—Hollywood).

Pal understood his market. Even if his films were, by the standards of the times, technical triumphs, his thematic development was standard. The coloring of the poster for his *War of the Worlds* makes clear that, like the Communists, the Martians are a "red menace." Despite the fact that both Wells's novel and Howard Koch's radio script have Martians marching across the Earth, Pal has them flying in saucers, like those flying over Mark Wicks's Martian capital of Sirapion. The tagline, "THE ORIGINAL INVASION!" clearly reminds the audience of more recent invasions past (Pearl Harbor?) and perhaps others soon to come (from Moscow?). Ray Bradbury showed us that, by destroying technology, we could possess a pastoral Mars, but Hollywood, based on film and ever more elaborate technical wizardry, possessed us with visions of a Mars unleashed at home.

Spacecraft: Us vs. Them

Hollywood works with images that its audiences understand. (See color illustrations at the center of the book.) When it comes to spacecraft, the code is clear: we send pointy, phallic rockets up at them and they send round, yonic saucers down at us. (By the by, craft that are built and travel endlessly between planets, like the *Enterprise* of *Star Trek*, rather than start on one planet or land on another, combine pointy and round components in a self-contained androgyny.) According to the standard symbolism, we penetrate the dark reaches of outer space to spread our seed on new planets while those planets, older than we, come down to smother us. The conflict in its most raw form is not simply that between essentially male Earth people (regardless of actual gender) and essentially female aliens (regardless of actual gender), but that between adolescent males and mature women. The symbolism is vivid in Robert Wise's classic film, *The Day The Earth Stood Still* (1951).

In the movie, the saucer is confronted by soldiers who gawk as an androgynous robot emerges from the alien craft. In outline, it resembles the breast of a recumbent woman. What will the aliens want of us? The soldiers, by the side of a tank, advance hesitantly and aim their puny-seeming can-

Photofest

non from afar at the slit in the alien saucer, symbolism that requires no verbal explanation. The shuffling of the unsure soldiers tells it all.

In the story, one main character is a boy who violates his single-mother's orders so he can spy on the doings on the Washington Mall, where the saucer has landed. The aliens, like scolding nannies, have come, alerted by our atomic explosions, to tell us we must cease pursuing such dangerous technologies or else, for the good of the rest of the universe, we will be punished. Only the mother, involved because of her son, succeeds in making the message truly heard by the authorities. In other words, the space opera allows rambunctious immature maleness to somehow both defy and support female rule.

Where the males win, as in *Earth vs. the Flying Saucers* (1956), the costs matter. (See color illustration at the center of the book.) Washington does withstand this attack from Mars, but, in the process, gets its Washington Monument sliced off. In *Rocketship X-M* (1950), the males don't deserve to win. The "M" of the title refers to "Moon," but a malfunction of their technology and a miscalculation of their fuel supply send them to Mars where

Amazing Stories, vol. 2, no.5 1927

they find a horrifying object lesson, a once strong civilization reduced to caveman status by their own atomic war. Two of the Earthmen die on Mars and the remainder, their rocket fuel prematurely spent, crash on their return to Earth.

The sexual iconography of spacecraft, and of Martians, begins before Hollywood takes it up and extends across media. Hugo Gernsback began reprinting Wells's *The War of the Worlds* about a year and a half after Gernsback named the genre of science fiction and created the first science fiction magazine. The cover of the August, 1927, issue of *Amazing Stories* depicts the Martian fighting machines, rather than their spherical meteor-like landing craft, but the shape of their machines is equally obvious; and, of course, every individually identifiable human in the path of these terrors is male.

Photofest

Off to Camp

One person's terror, of course, is another's turn on. In films like *Devil Girl From Mars*, a host of charged images and subconscious fears are handled with a camp irony so broad and a budget so low that one can only suppose that the implicit exchanges of knowing winks between the makers and the viewers of the film signal a complicit admission that those fears may be real after all. Without some underlying psychological engagement, how could anyone sit through a movie so badly made that a supposedly awe-inspiring robot of death is obviously a man with a gray, painted cardboard box loosely placed over his head? The special effects consist of models of a saucer and an aircraft "flying" on wires and a zap "ray" with which the robot makes lonely hilltop trees "disappear," to be replaced with the wisps rising from a tin smoke pot. In this 1954 British production, it is not unreasonable to see the racily black-clad "devil girl" not only as a dominatrix but a neo-Nazi. Atomic sex war on Mars has left the women winners in need of male "seed," so the

Photofest

title character, Nyah, invades Earth to snatch some studs. Of course, like any true low-budget film Nazi, Nyah spends most of her time decrying the local specimens as beneath consideration. Not that that would stop her from stocking up.

The plot, set almost entirely in and near a lonely pub in the Scottish Highlands, concerns a group of people held by Nyah until her atom-powered saucer can be repaired. One of the humans just happens to be a man who has escaped from prison. His crime? Wife killing. And he was guilty. But in the end, he manages to get himself taken onto the saucer and sabotages the reactor, blowing up Nyah, saving the Earth from future invasion, and proving once again that sometimes killing a woman can be a good thing. Mars, in other words, becomes a domain in which our perversions make us heroic. This film is just as antiscientific as the writings of Ray Bradbury, but offers a much Grimmer fairyland.

Back in the United States, Hollywood, too, exploited Mars as a resource for the silly indulgence of sexual fantasy. Consider *Abbott and Costello Go To Mars* (1953). Like the short cartoons with Marvin the Martian, this A & C adventure represents but one episode in a series of films that the audience already knows and expects to find funny. The tagline "It's Too Wild for One World" along with Abbott's face of expectant terror and Costello's infant dubiety, suggests that the comedy will be "out of this world" funny while the plot takes our familiar clowns literally out of this world. Indeed, in other posters, the tagline reads, "They're Out Of This World On A Misguided Missile." In a period when every image of a rocket reminds the public of an international arms race, embracing more than "one world" in laughter offers a moment of pointed respite. How appealing to be able to follow a *mis*guided missile for a while, phallic fun for the whole family. Just as the poster for *Devil Girl From Mars* was dominated by a dominatrix and her saucer, so here our male heroes ride the rocket with scantily clad starlets just waiting seductively in space.

The film begins with Orville (Costello), a school janitor, getting trapped in the delivery truck of Lester (Abbott) on his way to the space agency. They accidentally launch themselves in a rocket destined for Mars, but in their hands it arrives in New Orleans during Mardi Gras where A & C mistake celebrants in outlandish costumes for aliens. Soon enough a bank-robbing duo, fleeing the cops, commandeer A, C, and their rocket. They take off again, this time arriving on Venus which (goddess of love, right?) turns out to be a civilization composed entirely of women.

These two films represent cultural uses of Mars that become possible in the clutches of mass art in the 1950s. Unlike the journalist narrator in Wells's novel or the military officer protagonist in Burroughs' or the erstwhile governor authorial mouthpiece in Bradbury's, the heroes in both these films are working stiffs at best. Mars itself appears in neither film, but the mere conjuration of its name lures an audience and suggests a background of danger that comfortably translates into mild fantasies both perverse and quasi-pornographic. In feature films like these, and in cartoon shorts like those of Marvin the Martian, Mars becomes a charged toy for us to enjoy at our leisure. Welcome to camp.

Chemosphere

Courtesy Anaheim Public Library

Nowhere was America more playful than in southern California, the home of Hollywood itself. In 1949, John Lautner (1911–1994), a Michigan-born architect who had studied with Frank Lloyd Wright (1867–1959), designed a Hollywood coffee shop with soaring, "space-age" windows and façade. "Googie's Coffee Shop" (now demolished) gave its name to an audacious subgenre of roadside architecture, like the Satellite Shopland, a mall in Anaheim, California, with a neon false-Sputnik logo atop a quasi-launch tower, and the Donut Hole doughnut shop where customers drove their cars in through one giant, upright half doughnut arch and out another after picking up their treats. In less exaggerated form, the style has penetrated most modern cityscapes around the world in the twin arches of McDonald's Restaurants and, arguably, the Nike swoosh that adorns sporting goods.

Long before the taming of these futuristic shapes, the movement had a separate identity. It became known as "googie" architecture and Lautner became known as its founding father. In truth, some of the most persistently impressive examples of googie were by Lautner, but many of his works, like the sensuous Arango House overlooking Acapulco Bay, do not make us

Courtesy James Mitchell

smile; they simply stun. One work that combines both the space-age kitsch of googie and Lautner's pioneering architectural ideas is his most famous work, the Malin Residence in the Hollywood Hills, otherwise known as the Chemosphere.

As whimsical at the Chemosphere may seem at first, this 1960 building brilliantly uses a single, central column to exploit what had been considered an unbuildable site. The 1961 *Encyclopedia Britannica* called it "the most modern home built in the world." Charles Moore and colleagues, in a 1984 book about Los Angeles architecture, note that

> The only things that touch the impossible terrain are long stairs, a small cable car and that one column, which carries the utility lines as well as the entire structural load; floor-to-ceiling windows, in a continuous band, provide panoramic views while, at night from outside, their glow combines with red and yellow spotlights from below to hasten many impressionable newcomers toward the conviction that *The War of the Worlds* has begun.

Photofest

After campifying the images of Mars, the next logical step was to inhabit them. And if only a few could inhabit them today, in the future they would house Middle America itself. At that point, the saucer would not be a craft the Martians take to Earth but a craft we Earth people, having absorbed the many lessons of Mars, make part of our daily, Earthly routine (see photo from *The Jetsons* television series). From 1962–1963, that happy vision lived on America's television sets.

An Ace Book, a Division of The
Penguin Group (USA) Inc.

Robert A. Heinlein:
The Martian Savior

In 1961, between the building of the Chemosphere and the launching of *The Jetsons*, Robert A. Heinlein (1907–1988) published *Stranger In a Strange Land*. Although the dust-jacket claim that this is the "most famous science fiction novel ever written" exaggerates, the claim that it had grown from "cult favorite to bestseller to a classic in a few short years" does not. For a decade or more, it seemed that everyone knew this book, whether they had read it or not.

The title, according to the Bible (Exodus 2:22), is the origin of the name Gershom, a son of Moses born during the Hebrew captivity in Egypt who is said to become the progenitor of a line of priests. The title also refers to the situation of Heinlein's protagonist, Michael (meaning, in Hebrew, "Who is like God?") Valentine (the saint of our hearts) Smith (one who forges). Mike, the sole offspring and survivor of a failed international colonial expedition to Mars, had been raised by Martians, ancient and wise creatures the existence of which was and remained unknown to those on Earth. When Mike is discovered years later by a subsequent expedition from Earth, he seems to be alone. He is returned to Earth where, by a quirk of fate, he has legal title to all of Mars, its only native son and yet citizen of the progenitor states of the Earth's world state. Despite owning vast tracts that would make any of Bradbury's crass characters salivate, on Earth Mike seems nearly powerless (raised in that low gravity) and almost fatally naïve. He is put in a hospital for protection and extensive therapy to accommodate to Earth's culture and gravity. Political thugs seek him. Gillian, a sympathetic nurse, saves him by packing the slender, light-weight Mike into a suitcase and carrying him to her home where the first thing she does is bathe him. She is a nurse, after all, and the suitcase was dusty. Unbeknownst to Gillian, on Mars, with its arid conditions, committed interpersonal communion comes from sharing a thimbleful of water. And she has placed Mike in a whole tub. In his mind, he takes her instantly to be his "water brother." There is a knock at the door. Gillian leaves Mike in the tub. Gillian opens the door. The thugs bull their way in, one waving a gun, demanding to know Mike's whereabouts. As Gillian stalls, Mike feebly comes to the door and "groks" that these are bad people. He thinks one and then the shocked other into another dimension. The no-longer-threatening gun clatters to the floor, no hand left to hold it. Suddenly the apparently weakest character in the novel is revealed as the strongest.

The connection between Mike and Jesus is emblazoned on the novel's dust jacket. Jubal Harshaw, Heinlein's obvious mouthpiece, soon after this failed home invasion, scolds Gillian for even clothing Mike.

> "You're forcing on him your own narrow-minded, middle-class, Bible Belt morality."

> "I am not! I'm simply teaching him necessary customs."

> "Customs, morals—is there a difference? Woman, here, by the grace of God and an inside straight, we have a personality untouched by the psychotic taboos of our tribe—and *you* want to turn him into a copy of every fourth-rate conformist in this frightened land! Why not go whole hog? Get him a brief case."

In the course of the novel, Mike has many adventures, but the key point is that he spreads the wisdom he brings from his Martian "nest." He establishes a philosophy (a.k.a. religion) based on water, communion, peace, and the right knowledge of people and the world, a knowledge that gives gentle but implacable power. To know thus deeply is to "grok." The *American Heritage Dictionary* defines Heinlein's addition to our language as "To understand profoundly through intuition or empathy." At the end of the novel, Mike "discorporates," but to his followers he has merely moved on. There is a better somewhere ahead. By the 1960s, this is something that we can grok from Mars.

Enter NASA

Courtesy NASA/JPL–Caltech, P7875A

On July 14, 1965, for the first time in history, humans, using our machines, managed to take close-up pictures of Mars. The Soviet Union had been trying since 1960. Neither of the Soviet launches in October of that year even reached Earth orbit, which was comforting for the United States since the prize was not so much scientific as political. Ever since the atomic bombing of Hiroshima, on August 6, 1945, the U.S.S.R. and the United States had been locked in a simultaneous arms race and space race, the two competitions inextricable because both depended on improving delivery systems and each projected national power. While the United States beat the U.S.S.R. to the first atomic bomb (1945) and to the first hydrogen bomb (November 1, 1952), on October 4, 1957, with the launching of Sputnik, the Earth's first artificial satellite, the U.S.S.R. leapt ahead. America was worried, and the Soviets kept pressing. The race to Mars was war in another form.

In October, 1962, the Soviets only achieved Earth orbit with a would-be Mars mission, while a Soviet vehicle launched in November of that year got part way to Mars but its radio failed, so no one knows what happened thereafter. The Soviets' fifth Mars mission, later that same November, again only achieved Earth orbit. That same month, the United States finally launched its first Mars mission: Earth orbit only. Two years later, in November, 1964, the United States launched three missions, and Mariner 4, the middle one, finally succeeded. NASA, the United States National Aeronautics and Space Administration, had beaten the Russkies to Mars . . . for the sake of humanity, of course.

NASA's current chronicle of this mission is telling:

> After an eight-month voyage to Mars, Mariner 4 makes the first flyby of the red planet, becoming the first spacecraft to take close-up photographs of another planet. The images show lunar-type impact craters, some of them touched with frost in the chill Martian evening. A television camera onboard takes 22 pictures, covering about 1 percent of the planet. Initially stored on a 4-track tape recorder, these pictures take four days to transmit back to Earth.

NASA's language offers a subtle mix of the scientific (specific numbers like "22" and "1 percent") and the romantic ("touched with frost in the chill Martian evening"). While Mars is first called "the Red Planet," a phrase Homer might have used in ancient Greece, soon "the images show *lunar*-type impact craters." In other words, thanks to NASA, Mars is suddenly much closer to Earth than it was; it is more like the Moon. Our technology may not be speedy yet, but it works, we have time, we are ahead. By today's standards, the few, grainy, black-and-white images are primitive, but at the time they put Mars fully within reach. (What is left wonderfully unsaid is that there are no signs of canals.) Now it was just a matter of time.

My Favorite Martian

Photofest

If Mars is so close to Earth, why wait until the time of the Jetsons to meet our neighbors? From September 1963 to September 1966, Ray Walston (1914–2001) played the part of Uncle Martin on a popular television prime time situation comedy called *My Favorite Martian*. In their directory of television shows broadcast from 1946 to 1992, Tim Brooks and Earle Marsh give a good capsule description:

> On the way to cover an assignment for his paper, *The Los Angeles Sun*,* reporter Tim O'Hara stumbles upon a Martian whose one-man ship had crashed on Earth.ᶜ Tim took the dazed Martian back to his rooming house to help him recuperate, while thinking of the fantastic story he would be able to present to his boss,✳ Mr. Burns,✸ about his find. The Martian, however, looked

---

*ESR: We are back in Los Angeles, where television was being filmed and one could look up at the Chemosphere.
ᶜESR: This Martian represents no threat to Earth at all.
✳ESR: Tim's first thought, like that of so many other characters, is to use Mars for personal profit.
✸ESR: As usual, the ranking capitalist is inherently destructive, or so his name suggests.

human, spoke English,✱ and refused to admit to anyone but Tim what he was.ᶜ Tim befriended him, passed him off as his uncle, and had many an interesting adventure with the stranded alien. Uncle Martin had little retractable antennae, could make himself invisible, was telepathic, could move objects just by pointing at them, and had a vast storehouse of advanced technological knowledge.✶ While he was trying to fix his ship he stayed with Tim in Mrs. Brown's❋ rooming house. During the first season, Mrs. Brown's teenage daughter Angelaᶜ was a cast regular. The following year policeman Bill Brennan joined the cast as Mrs. Brown's boyfriend, a threat to Uncle Martin on two counts—not only was he always a potential discoverer of the Martian's true identity,✱ but Uncle Martin had become romantically interested in Mrs. Brown himself and looked upon Brennan as a rival for her affections.✱

✱ESR: In other words, Martians *are* our cousins.

ᶜESR: Suddenly the Martian has a role much like that of a standard fairy godmother, but in science fiction, we find a manlier godfather. Will workaday Tim, living in someone else's home, ever make it on his own?

✶ESR: Martin shows standard Martian features of both biological and technological evolution, yet, visiting us every week in our living rooms, he was domesticated.

❋ESR: Mrs. Brown: Earth tones?

ᶜESR: The daughters of the Earth are angels.

✱ESR: Organized government, like that in *The Day the Earth Stood Still*, remains the real threat, rather than the Martian. We have not been invaded, but in the era of the space race, we would act as if we had been if we only realized that there is a Martian among us.

✱ESR: Again we find the Martian highlighting sexual competition. However, this competition is clearly benign. One can no more imagine sweet and wry Uncle Martin forcing himself on Mrs. Brown than his hooking up with Nyah, the neo-Nazi "devil girl from Mars."

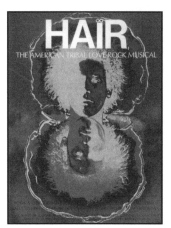

Photofest

The Age of Aquarius

From the ancient Babylonians onward, the zodiacal month we know as Aquarius (named after the constellation of the water-bearer) has been important for renewal. Aquarius falls from our mid-January to mid-February, a time when rains came to the parched Middle East. In some astrological systems, there exists the notion of a "great year," about 25,000 ordinary years in duration. In that scheme, the Earth moved into the "age of Aquarius" in the early nineteenth century. But Madame (Helena) Blavatsky (1831–1891), a Russian-born spiritualist and cofounder of the Theosophical Movement in London, made popular the notion that people of many religions and all nations need to cooperate in their own behaviors and spirituality to bring about a true Age of Aquarius, a new era of peace, ease, and love. Her followers and others predicted this renewal at some time in the last quarter of the twentieth century. The spiritual interests of the American hippies and others in the 1960s resonated with this promise of imminent peace, a powerful promise at a period when a substantial segment of America felt or demonstrated varying degrees of resistance to the war in Viet Nam.

In 1966, *Hair* burst onto Broadway. This antiwar, antiestablishment, exuberant hippie musical by Gerome Ragni (lyrics) and Galt McDermot (music) focused mainly on the lives of the resisters, the Flower Children, those who wanted to "make love, not war." The show shocked and delighted contemporary audiences in part by a famous "nude scene" in which, in fact, all the cast members, in quite dim light, were suddenly revealed stock-still and naked for a brief moment before the intermission curtain rang down. The gentility of that notorious moment helped make *Hair* not only a commercial but a cultural success.

Among the many memorable songs in *Hair* was one that became a counter-cultural anthem. "Aquarius" begins with a stanza that became its refrain, promising that when Jupiter and Mars align "peace will guide the planets / And love will steer the stars." That glorious moment ushers in "the age of Aquarius," with the ringing word Aquarius repeated again and again.

These astrological lyrics reflect a modern social critique, one perfectly in keeping not only with hippie values but with the sharp-edged 1961 warning by outgoing American President Dwight Eisenhower (1890–1969): "In the councils of government, we must guard against the acquisition of unwarranted influence, whether sought or unsought, by the military-industrial complex." This from a former general. True, Mars is the god of war, but Jupiter (the whole capitalist system) outranks him. Jupiter, father of Mars, rules the Olympians. We must help these two to "align"—like a chiropractor easing spinal tension—so that "peace will guide the planets" including, of course, the Earth.

The lyrics of "Aquarius" not only repeat the necessity of Jupiter aligning with Mars but employ recurring water imagery, from the obvious ("to be the bearers of the water") to the allusive (the repetition of the word "Aquarius" itself) to the subtle ("crystal revelations"). What many may have noticed in the late 1960s, but might not today, is that this use of water imagery for binding people together, based on the alignment of Mars with the overall dominant system, loudly echoes the plot of Heinlein's *Stranger in a Strange Land*, at that very period "a cult favorite and classic." Just as Heinlein's Mike discorporates to save us, one of the hippies in *Hair* goes off to war and dies. Like a deadly serious Uncle Martin, the sometimes invisible favorite Martian, someone, this anthem prays, will be there to "care for us."

The Face on Mars

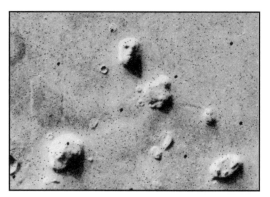

On July 31, 1976, NASA, probably inadvertently, fueled the imaginations of multitudes hoping to find on Mars answers to humanity's persistent questions. The picture above accompanied the following press release:

> NATIONAL AERONAUTICS AND SPACE ADMINISTRATION VIKING NEWS CENTER PASADENA, CALIFORNIA (213) 354-6000 Viking I-61 P-17384 (35A72). July 31, 1976.—This picture is one of many taken in the northern latitudes of Mars by the Viking 1 Orbiter in search of a landing site for Viking 2. The picture shows eroded mesa-like landforms. The huge rock formation in the center, which resembles a human head, is formed by shadows giving the illusion of eyes, nose and mouth. The feature is 1.5 kilometers (one mile) across, with the sun angle at approximately 20 degrees. The speckled appearance of the image is due to bit errors, emphasized by enlargement of the photo. The picture was taken on July 25 from a range of 1873 kilometers (1162 miles). Viking 2 will arrive in Mars orbit next Saturday (August 7) with a landing scheduled for early September.

National Park Service

As scientific as this may seem, the press release invites seeing a Face on Mars as readily as many people see a Man in the Moon. More subtly, the use of the term "mesa-like" recalls, in the words of the *Oxford English Dictionary* definition for mesa, ". . . a flat-topped hill or plateau of rock with one or more steep sides, usually rising abruptly from a surrounding plain and common in the arid and semi-arid areas of the United States." In other words, Arizona, Percival Lowell and Edgar Rice Burroughs country.

Burroughs grew up and spent substantial time as a traveling salesman in the Midwest. At numerous Midwestern sites, including what is now Effigy Mounds National Monument near Harper's Ferry, Iowa, not very far from Chicago, one can find burial mounds in the form of effigies, typically of animals, that disclose their shapes only from above, as in the case of a bear-shaped mound with its revealing rock border created by Park Service personnel.

Even if Burroughs had no idea of the connection between such aerial effigies and Native Americans, and so he could not place such a Face on Mars as a native artifact, the NASA press release copywriter must have been aware of such popular phenomena as the enormous Nazca Plain markings, ancient drawings in the desert 200 miles south of Lima, Peru. These extensive designs, discovered by the West only in the 1930s, were formed by moving

Fortean Picture Library

dark material off light in such a way that a pattern is formed in the earth. However, the pattern is so large that it is visible only from the air. After their accidental discovery from overflying aircraft, authors like Erich von Däniken (1935–) suggested that these designs were created under the direction of aliens to guide them in their landings on Earth. If aliens could create signals here on Earth, surely they could create them on their own planet.

In 2000, the Hollywood film *Mission to Mars* exploited the premise that the Face on Mars was not only a sign but an enormous building with wonderful, heavenly secrets waiting to be revealed. Reviews at the time noted the wide popularity in books, magazines, and television shows of the idea that the Face on Mars reflected an ancient culture signaling to someone. The movie was advertised with the tagline, "Let There Be Life." The extended tagline: "For centuries, we've searched for the origin of life on Earth. . . . We've been looking on the wrong planet."

But the Face on Mars may be a hope hard to entertain for much more than the quarter-century it has already enjoyed. In April 2001, NASA's Mars Orbiter Camera sent back a much finer-grained picture of the same feature. It looks considerably less like a face . . . to some people.

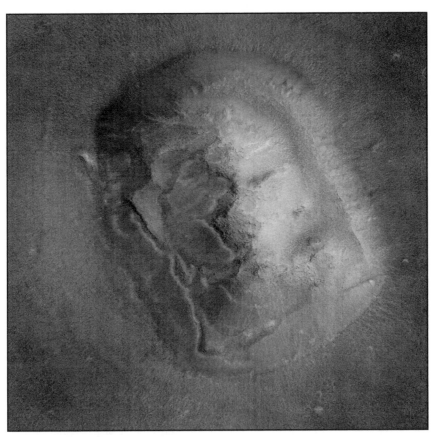

NASA Jet Propulsion Laboratory, Malin Space Science Systems

Mars Attacks!

Apparently we have lived with our images of Mars so long that we can thoroughly ironize them. In the poster for the broadly parodic 1996 movie called *Mars Attacks!* (see color illustration at the center of the book) we see the evolved big-brain alien with a nasty, death-dealing skeletal smile; the sexual competition in the pointy-breasted model (who turns out to be a Martian remodeled like a pod person to appear human); and the fairy-tale punishment of a woman melded with a dog. The backdrop to this is the Red Planet, which is releasing saucers speeding our way and causing flaming destruction. In other posters, the filmmakers let us know that this end-of-the-century spoof intends to satirize not only our images but the earnest warnings of earlier science fiction movies like *Earth vs. the Flying Saucers*. In *Mars Attacks!*, too, Washington is covered by descending saucers. But to the blasé 1990s, this is all just hilarious.

Leaving Earth Behind

In this comic view of the Milky Way galaxy, found on posters and T-shirts, we seek wayfinding not to reach Mars but some destination so impossibly more distant that even our own Sol, not to mention planets like Mars and Earth, is indistinguishably close by comparison. And yet, given the reference to such mundane and ubiquitous wayfinding signs, the kind we see in thoroughly humanized environments like large city parks, this satiric guide suggests that wherever any other planet might be, it will be easy to reach.

Mars Today

As a NASA composite photo makes clear (see color illustration at the center of the book), Mars is no longer to be *imagined* as Arizona; it is to be *seen* as Arizona, which puts it right in most Americans' metaphorical backyard. But note the typically subtle rhetoric in this excerpt from the accompanying NASA press release.

> The Twin Peaks are modest-size hills to the southwest of the Mars Pathfinder landing site. They were discovered on the first panoramas taken by the IMP camera on the 4th of July, 1997, and subsequently identified in Viking Orbiter images taken over 20 years ago. The peaks are approximately 30–35 meters (~100 feet) tall. North Twin is approximately 860 meters (2800 feet) from the lander, and South Twin is about a kilometer away (3300 feet). The scene includes bouldery ridges and swales or "hummocks" of flood debris that range from a few tens of meters away from the lander to the distance of the South Twin Peak.

Twin Peaks is a perfectly good name for this Independence Day discovery, but in 1997 most American adults would have thought of "Twin Peaks" as the name of a then famously bizarre television series (1990–1991) that begins with an unsolved mystery and takes ever more surreal twists. Here NASA wants to remind us of mystery and yet tame it. The picture shows what look like enormous mountains rising above a distant horizon, even as the texts calls these "modest-size hills" and provides the numbers to support that. If ever there were mystery here, NASA wants to reassure you that these hills really appeared in photos as long ago as twenty years. So there's nothing to worry about or even be excited about. It's just Mars.

Phobos and Deimos Today

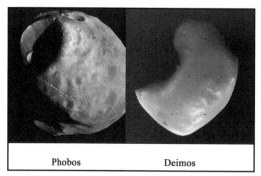

| Phobos | Deimos |

NASA, NSSDC

Thanks to pictures like these from NASA's Viking Orbiter, Fear and Terror have been reduced to pebbles. The biggest mystery may be in explaining the contrast in surface markings between grooved and cratered Phobos and smooth Deimos. Of course, neither is round, so perhaps, unlike Earth's moon, they began as asteroids and were captured by Mars' gravity, as some astronomers have speculated. Maybe that will prove to be the biggest mystery. Either way, they don't send chills up our spines. These are questions that rightly matter to science, but if you loose the chill, you risk losing public enthusiasm and with it public funding.

Men Are From Mars . . .

Courtesy Cartoonstock.com; drawing by Chris Grosz

In a world in which "Mars" has become "just Mars," the cultural meanings of the planet revert to the metaphors that have been the least common denominator of the planet throughout history. John Gray's 1993 bestselling self-help book capitalized in its very title on the notion that Mars and Venus have different aims, different strengths, different social relations, but both are powerful. The book's tagline suggests that "communication" has been a problem (as one imagines it would be, given the distance between Mars and Venus) but that if you can fix that, you can "[get] what you want." In other words, despite the possibly communal implications of the word "relationships," the book promises individual success. "Mars" denotes sexual competition, as is so often the case, and the book will teach us how to win.

The simplicity and resonance of the title entered modern vocabularies. Nick Park, the award-winning animated film artist who created the Wallace and Gromit series about the adventures of a bemused, inventive bloke and his wisely tolerant dog, named one of their episodes "Men Are From Mars, Dogs Are From Pluto" (2002). (The reference here is not only to

Cover art work from Ron Robinson, *Cats are From Saturn . . . Dogs are From Pluto* (Ex Machina Publishing, Sioux Falls, S.D.: 1998)

Gray, but to Walt Disney who gave Mickey Mouse a dog companion named Pluto in 1930, the year of the discovery of the planet Pluto.) And, thanks to Mars and Gray, the whole realm of planetary metaphor opened up, sometimes even leaving Mars itself far behind, as in a short book of humor Ron Robinson published in 1998.

The NASAfication of Mars

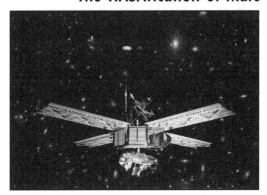

NASA/JPL

According to the *Oxford English Dictionary*, since the early fifteenth century, the word "nominal" (from the Latin word for "name") in English grammar has meant "[o]f the nature of, pertaining to, a noun or nouns." By the early seventeenth century "nominal" had developed its most common current use, "in name only, in distinction to *real* or *actual*; merely named, stated, or expressed, without reference to reality or fact." (Example: He had nominal control of the situation, but in fact unseen forces controlled him.) In 1966, NASA turned that use around completely by inaugurating a new use of "nominal" as "meaning within prescribed limits; anything from 'perfect' to acceptable."

In 1969, NASA launched two Mars missions, Mariner 6 (February 24, 1969) and Mariner 7 (March 27, 1969). On July 31, 1969, Mariner 6 performed a "flyby" pointing its telescopes at the Red Planet and its antennae at Earth. ("Flyby" was coined in the 1950s by military jet aircraft pilots, a group from which NASA drew heavily in recruiting astronauts and other personnel. It entered the language at large trailing with it connotations of danger, adventure and control. Dictionary quotation from

1953: "On each day of the show there will be spectacular aerial flybys of jet planes.") The Mariner 6 flyby sent us 75 photos. Its twin, Mariner 7, made its flyby on August 5, 1969 and sent back 125 photos. Together, they covered about 20 percent of the surface of Mars, including some areas of enormous interest.

Because so many people, often billions, are watching, and because the activities are often historic firsts, both the successes and the failures of NASA have important cultural consequences. One Mariner flyby would have been a scientific triumph; two made NASA seem like an assembly line. To the public, each Mariner mission might have been "an adventure," but NASA, aiming to project its engineering prowess and reassure the public that it is safe to fund space exploration, called these missions "nominal." While "flyby" may have set the blood racing for the NASA "in" crowd who associated the term with heroes who had "the right stuff," to most Americans, the term was new and just meant "flying by." In fact, the report of a "nominal" flyby carried the tin ring of a not-so-real flyby. This effect was subtle, and unwanted by NASA, but real nonetheless.

To maintain its public attractiveness, NASA exploits two fundamentally important but fundamentally contradictory rhetorical positions. On one hand, NASA wants to be daring and pioneering; on the other, NASA wants to be trustworthy and reliable. Sending teacher Christa McAuliffe aloft as "the first teacher in space" attempted to join those hands. As "the first," she was a daring pioneer; as a passenger on the space "shuttle," a vehicle as ordinary-sounding as a tow truck or ferry, she would demonstrate that the system was trustworthy and reliable. Tragically, shortly after launch, she and the entire crew with which she flew died quite shockingly when the shuttle *Challenger* exploded on January 28, 1986.

Some said that *Challenger* had carried a teacher prematurely, that the system obviously was not yet fully reliable. Others said the system had become so reliable that the public had lost interest, so the teacher was recruited as a public relations gambit. Either way, NASA lost. In fact, as with most public moves by NASA, both aims and both criticisms had truth: The system was very reliable, but not as reliable as claimed; the teacher was a pioneer as well as a public relations ploy. And what, after all, would be so wrong with that? Didn't Ms. McAuliffe compete for her place, accept the dangers, and give her life not simply for a personal thrill but to inspire children everywhere? And she did inspire them, even in death.

NASA has experienced three terrible, headline-generating fatal accidents: the immolation of astronauts Gus Grissom, Roger Chaffee, and Ed White

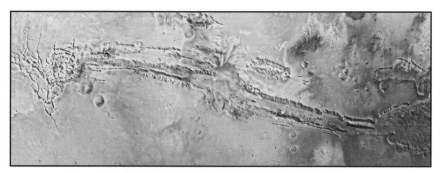

NASA/JPL/Arizona State University

when they were trapped in an Apollo space capsule being tested on Earth (January 27, 1967); the *Challenger* explosion; and the re-entry disintegration of the shuttle *Columbia* (February 1, 2003). But each time, NASA moved forward by publicly acknowledging that pioneering is dangerous and visibly applying to their investigation the best engineering expertise available. They also rose again on the spirits of their people. Before the Apollo disaster, Gus Grissom said, "If we die, we want people to accept it. We're in a risky business, and we hope that if anything happens to us it will not delay the program. The conquest of space is worth the risk of life." After the *Columbia* tragedy, the next trumpeted step in this conquest was, in typical NASA fashion, twin missions to Mars.

In the years after NASA quietly dispelled the myth of the Martian canals, it strove, consciously or not, to create new myths that accorded with Mars as a fit landscape for the human (and sometimes specifically American) pioneering spirit by offering pictures to the public, which NASA proudly and gratefully acknowledged had paid for them, and explanatory text that continued to attempt to marry the dangerous and the doable. Two of the key features of Mars now became Valles Marineris and the Olympus Mons.

NASA describes the Valles Marineris in these words:

> The Mariner Valley, also known as the Valles Marineris canyon system, appears in this mosaic of images from NASA's Viking spacecraft as a huge gouge across the red planet. This "Grand Canyon" of Mars is about 2500 miles long and up to 4 miles deep. By comparison, the Earth's Grand Canyon is less than 500 miles long and 1 mile deep.

NASA describes Olympus Mons (Mount Olympus, the home of the gods, including Mars) in these words:

> Olympus Mons on Mars is the largest volcano in the Solar System. Although three times higher than Earth's Mount Everest, Olympus Mons would not be difficult to climb because of the volcano's great breadth. Covering an area greater than the entire Hawaiian volcano chain, the slopes of Olympus Mons typically rise only a few degrees at a time. The low gravity of Mars combined with a relatively static surface crust allow volcanoes this large to build up over time. This representative-color image was taken last April [1998] by the Mars Global Surveyor spacecraft currently orbiting Mars.

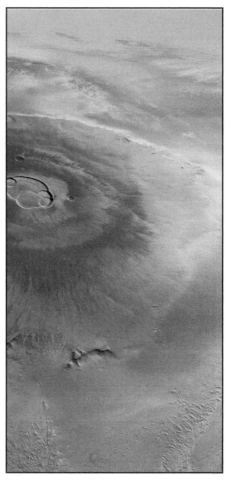

NASA/JPL/Malin Space Science Systems

Another press release accompanying the same image reiterates some of the key geological points and adds a meteorological description reminiscent of *The Martian Chronicles*:

> Taken on April 25, 1998, from a distance of about 900 kilometers (560 miles) above the surface, this wide-angle image of Olympus Mons captures the west side of the volcano on a cool, crisp winter morning. Olympus Mons is by far the tallest volcano in the solar system, rising higher than three Mount Everests and spanning the width of the entire Hawaiian island chain.

NASA–Marshall Space Flight Center

Well, who wouldn't want to go to Hawaii, or climb an easy triple-Everest, especially if you get to hike a quadruple–Grand Canyon along the way? It is clearly time for a vacation. On June 10, 2003, NASA launched the first of its two Mars Exploration Rovers.

Mars: The Eighth Continent

The land area of the Earth is approximately equal to the total surface of Mars.

The land area of Africa is about the same as the total surface of the Moon.

NASA/JPL, art by Corby Waste

This image, produced by the Jet Propulsion Laboratory, a prime NASA contractor, puts Earth, Moon, and Mars in perspective. It goes without saying—for JPL did not want to say it—that the distances are not drawn to scale. The Earth shows all the continents pushed together almost like an image of the ancient Pangea, the proto-continent that gave rise to our current seven. If this is the Earth, it is a prehistoric Earth, a prehuman Earth, an uncorrupted Earth, and yet a familiar Earth that we all know and love. Hence, if the Moon, the nearby little Moon, is like Africa, we Westerners recognize that it is somewhat less familiar to us, but it is part of our world, a place we could visit someday, a place worthy of our attention. And if that is so, how much more so Mars, which is larger, probably rich with who knows what unexploited resources, and not that much further away.

Marscape

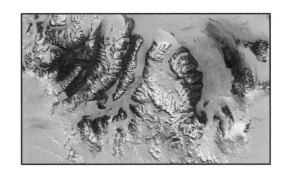

Image by Robert Simmon, based on data provided by the NASA GSFC Oceans and Ice Branch and the Landsat 7 Science Team

In *Victorians and the Machine* (1968), Herbert L. Sussman (1937–) points out that there is a pivotal period in the middle of the Victorian era, particularly visible in the language of the novels of Charles Dickens, in which we see the realm of the tenor (the "meaning") of metaphors and similes and the realm of the vehicle ("that which carries the meaning") change places. Where once one might have been most likely to say, "the camera is like an eye," one became more likely to say, "the eye is like a camera." That rhetorical exchange suggests a change in one's senses of familiarity. Where once the default understanding of the world was that it was natural, as people lived more in cities, worked in factories, and moved in street cars, the default understanding became that the world was artificial.

NASA works hard to make Mars seem Earthlike, although not so Earthlike as to undercut the adventure of going there. "A cool, crisp winter morning [on] Olympus Mons" has an undeniable appeal, one that, apparently, has penetrated the minds of many. In its February 3, 2003 issue, with the headline "Marscape: McMurdo Dry Valleys," *Time* magazine reflected on the

NASA, Ames Research Center

desolation at the bottom of the Earth (shown at the beginning of this chapter) in the following terms:

> That Antarctica is so little understood is not surprising, for it is a remote, otherworldly place that in many ways resembles early Mars more than contemporary Earth. And no place is more Martian in character than the McMurdo Dry Valleys, a wedge of rugged, rocky terrain stippled with ice-covered lakes and overhung by glaciers. No diminutive alpine plants cling to the slopes of these valleys. No rodents scurry amid the boulders and scree. No flies or mosquitoes whir through the air; no fish, mollusks or crustaceans dwell in the lakes and streams.

Here the tenor of the explanation is Antarctica, the vehicle Mars, suggesting the writer's sense that Mars is the more familiar locale. The same language that distances Antarctica as Martian embraces Mars as Earth-like. This follows perfectly the attitude expressed earlier in the May, 1991, issue of *Life*, *Time*'s corporate sibling. The cover bore the headline OUR NEXT HOME and showed only a color photograph of Mars with the subtitle, "Mars: Bringing a dead world to life."

Is the coincidence of the subject matter and the title of the magazine accidental? Perhaps. But perhaps not. The magazine, *Life*, was about *Life*, and the idea of life had already been planted, even by the U.S. government, on Mars, both in engineering plans and in artists' depictions, as shown in the NASA "historic" illustration on the previous page.

Terraforming Mars

The *Oxford English Dictionary* defines "terraforming" as "[t]he process of transforming a planet into one sufficiently similar to the earth to support terrestrial life." The first recorded use of the words is "1949 'W. STEWART' in *Astounding Sci. Fiction* Feb. 15-1: I've got the Martian industrial trust interested in an atomic furnace to make synthetic terraforming diamonds." W. Stewart was a pseudonym of Jack Williamson (1908–), a grand master of pulp science fiction. The novel in which "terraforming" first appeared was *Seetee Shock*, published serially in February, March, and April, 1949, in *Astounding*. "Seetee" came from "C.T.," "contraterrene" matter, what we usually call "antimatter." When antimatter and ordinary matter meet, there is an immediate, enormous conversion of mass into energy. *Seetee Shock* was the second of two serialized Seetee novels (the earlier was *Seetee Ship*, 1942–1943) in which this process is harnessed to make life possible even on the larger asteroids. These novels were so successful that they were reissued multiple times in book form and led Williamson to create a newspaper comic strip, *Beyond Mars*, that ran for three years (1952–1955) in the *New York Sunday News*, a comics-laden tabloid distributed throughout the United States. The notion of terraforming gained national currency.

Most planets, perhaps all, are not duplicate Earths. In pursuit of stories of far-off human exploration and habitation, science fiction has sometimes resorted to the remaking of people to fit their alien environments. Famous works in that vein are James Blish's *The Seedling Stars* (stories published in book form in 1957) and Frederik Pohl's *Man Plus* (1976). But in keeping with the positivism of science itself, the vast majority of science fictions that deal with the fit of human organism to alien environment have chosen to change not us but whole other worlds.

In science fiction, terraforming, one of the most grandiose engineering subjects imaginable, is sometimes treated utterly fancifully, as in Ray Bradbury's "Green Morning" section of *The Martian Chronicles*, and sometimes pursued on a comparatively small scale, as in Arthur C. Clarke's *The Sands of Mars* (1951). However, sometimes the subject is fully serious. Especially after the publication of some popular nonfiction books that touch on terraforming (like Carl Sagan's *The Cosmic Connection*, 1973, and Adrian Berry's *The Next Ten Thousand Years*, 1974), terraforming stories spread in audience and deepened in thoughtfulness. Today there exists a Mars Society dedicated to fostering the exploration of Mars and also another organization, Red Colony, that distinguishes itself from the Mars Society by being committed to inhabiting rather than merely exploring Mars. Red Colony, like most terraforming science fiction of the last thirty years, understands that brute mechanical force alone (moving planetary orbits, erecting huge sunlight-gathering orbital mirrors, and so on) probably will not do the trick. Often the missing piece is bioengineering on a global scale, "The Green Morning" made plausible.

Kim Stanley Robinson (1952–) has published the acknowledged masterpiece of terraforming, a massive trilogy consisting of *Red Mars* (1993), *Green Mars* (1994), and *Blue Mars* (1996). Robinson combines thoughtful, detailed analysis of biology, astronomy, chemistry, physics, and meteorology to provide the compellingly inventive backdrop for an exploration of the years-long human effort necessary to coax slow change out of a stable and alien world. And not just any world, but arid, light-weight, thinly atmosphered Mars. Yet despite the sensible extrapolation, the cover illustration of *Green Mars* (see color illustration at the center of the book) reminds us of our goal, a return on Mars to Eden, or at least to the New World of America, a pastoral hope that accords perfectly with the half-century earlier dream of Bradbury.

Robinson established a livable Mars so thoroughly that shortly after he began publishing his Mars trilogy, other writers presumed the success of terraforming. Karl Schroeder, for example, in *Ventus* (2000), set in the far future on a distant world in another solar system, has a character make a single offhand remark about Mars: "'You know what Mars is like—a hundred billion people stacked in pods like so much lumber dreaming their own universe into being while the physical infrastructure of the planet crumbles around them.'" In a post-Robinson writing world, Schroeder shows that (a) Mars is still the site of wish fulfillment and the field of fantasy; (b) Mars still represents an older version of Earth; (c) Mars is still decaying (as in Wells); (d) Mars is Bradbury's dream gone sour through its own success, like Wells's view of evolution, both biological and planetary, in *The Time Machine*.

But the end-of-time disappointments of fiction do not deter the real-life quest for the possible promises of Mars.

Red Rover

Although NASA, like Schroeder, certainly concerns itself with the universe beyond Mars, it also concerns itself with Mars. The government-provided artist's conception of one of the two identical Mars Exploration Rovers (see color illustration at the center of the book), issued well in advance of their first mission launch, shows a doughty device standing firmly on its six sturdy wheels. Its main headlike antenna mast is held proudly high. This is our rover ("Here Rover, here!"), a faithful, mechanical companion we seem to play with in Arizona, albeit by remote control. The tail looks perky, its horizontal surface staunch. The background reminds us of NASA's newly named Twin Peaks. The viewing angle makes it appear that Rover is somehow higher than the mountains themselves.

Time magazine bought into NASA's imagery. In a special section (November 18, 2002) called "Coolest Inventions 2002," *Time* presented the NASA image with the following one-paragraph story:

RED ROVER, RED ROVER

In 2004, the hottest car in the word will have a top speed of 10 ft. per min.—if, that is, the world happens to be Mars. NASA is currently testing two robotic rovers to send to the Red Planet in a mission set to launch next summer. The two solar-powered vehicles will travel up to 330 ft.—compared with Sojourner's 16 ft.—a day while using their nine cameras and three spectrometers to make scientific observations. First on the agenda? Looking for traces of water.

Inventor: NASA
Availability: January 2004
To Learn More: fido.jpl.nasa.gov

The title recalls a children's running game in which teams call out to each other "Red Rover, Red Rover, let [fill in the name] come over." What could be more simple, and joyful, than that? Note that NASA itself recognizes that "Rover" is traditionally a dog's name: The web address that offered more information begins with "fido," an even more traditional name for a dog. But this is our dog, a heroic dog, a wonder dog.

Until you turn from the artist's conception to the real thing. Even with its masts up, Rover is substantially smaller than any person in the room. Our conquering hero, looked at another way, is a Martian Lego lander.

Jet Propulsion Laboratory

Water on Mars

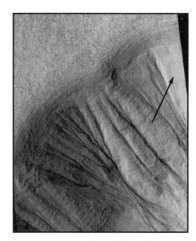

NASA/JPL/Malin Space Science Systems

The *New York Times*, on February 20, 2003, printed this picture under a headline that read:

PHOTOS BOLSTER IDEA OF WATER, AND POSSIBLY LIFE, ON MARS

A new theory and a revised interpretation of earlier observations have bolstered the idea that Mars has more water than previously thought and encouraged speculation about the possibility of life on the planet.

The new theory, identifying melting snow deposits as the likely cause of the many deep gullies on the planet, was announced yesterday at a briefing at the National Aeronautics and Space Administration in Washington. Photography from the Mars Odyssey spacecraft, orbiting the planet, inspired the suspected relation between melting snow and the gullies. A week ago, other scientists reported new studies showing that both Martian polar regions are capped almost entirely with ice, enlarging the planet's known reservoir of water. Until recently, the huge southern cap was thought to be predominantly, if not entirely, frozen carbon dioxide, or dry ice.

The arrow in the picture points to snowpack that is supposed seasonally to generate the meltwater that forms the gullies. It is doubtless only a coincidence that the first scientist mentioned as advancing this theory, Dr. Philip Christenson, is "a planetary scientist at Arizona State University."

Given these photos, the Rovers had real hope of finding water and, as other scientists quoted in the article affirmed, where there is this much water, there may once have been, or even be, life. Later, follow-up articles, suggested that soon enough that life might even be, with a bit of clever nudging, us.

NASA had publicly entered the business of terraforming.

Land of Spirit and Opportunity

NASA–KSC

On April 4, 2001, NASA launched its first Mars mission of the twenty-first century. The Mars Odyssey is an orbiter that took up its station circling Mars and began conducting "science mapping." One reason for that mapping was to locate the best landing sites for the twin Martian Exploration Rovers (MERs). On June 9, 2003, the day before the first MER finally launched, the *New York Times* published a photograph of the waiting rocket with the simple headline, "Weather Delays Launching of Mars Probe." The word "probe" in the headline may alert us to the iconography of an enormous rocket, surrounded by smaller yet equally penetrating shapes, towering over the little men at its base. Seeing the probe pointed powerfully upward, it seems almost like a reverse variation on *The Day the Earth Stood Still* where this time the men really do have the big guns.

The accompanying story, too, concerns much more than weather.

CAPE CANAVERAL, Fla., June 8—A forecast of high winds and severe storms forced NASA to delay today's launching of the first of two probes designed to land and look for water on Mars.

Weather permitting, the space agency plans to try again on Monday.

The two probes, robots on wheels, will look for evidence that water played a role in the planet's geological history.

The probe that was to be launched today, Spirit, and an identical probe to be launched this month, Opportunity, weigh about 400 pounds each and carry scientific instruments to study the composition of the Martian soil and rocks. The tools include milling cutters to drill the outer layers of rocks and boulders, and microscopes to inspect their interior.

The postponement of the launching came two and a half hours after the National Aeronautics and Space Administration officially announced the name of the spacecraft. A nine-year-old girl from Scottsdale, Ariz., who was born in Siberia and adopted as a toddler, submitted the winning names for both rovers.

"I used to live in the orphanage. It was dark and cold and lonely," said Sofi Collis, reading her essay at a news conference.

"At night, I looked up at the sparkly sky. I felt better. I dreamed I could fly there. In America, I can make all my dreams come true. Thank you for the 'Spirit' and the 'Opportunity.'"

Sofi's choice for the probes' names was picked from almost 10,000 entries submitted by United States schoolchildren. The contest was sponsored by the Lego Company.

The story does not tell us how Sofi's essay itself was chosen, but her idea of calling the rovers "Spirit" and "Opportunity" could not ring more truly as anthems Americans want to hear sung of their nation. This probe, then, is constructed to bring American values to Mars, to give America a new landscape to shape.

Sofi herself seems to be living another Horatio Alger story. Sofi, like Little Orphan Annie, has been snatched from poverty to the protection of moneyed America only to rise and demonstrate by winning a national competition herself that this faith was well placed. Her native land is Siberia, synonymous with inhuman conditions and ice. But just as the ice of Siberia could melt for Sofi, so the snowpack of Mars could provide water, carried to colonists who need it. This mimics the people of Scottsdale who live in the Southwestern desert and drink water brought by massive aqueducts. The entire contest enterprise, we learn, is sponsored by a commercial concern, a toy company at that. That the company is Danish, rather than American, suggests how great is the power of American culture to exploit all it surveys, including both the idea and the actuality of Mars. However, this international effort to launch the concrete realization of a child's dream also suggests something much grander: the enduring, universal, imaginative power pulsing in the Red Planet Mars.

August, 2003

On August 27, 2003, Mars came closer to the Earth (34.646 million miles) than it had been in 59,619 years and closer than it would be again until August 28, 2287. NASA trained its Hubble Space Telescope on the planet and produced the best pictures we're likely to get without actually going there. (See color illustration at the center of the book.) Of course, we were going there, on the wings of Spirit and Opportunity. After imagining the Red One since the dawn of human history, as we approached it, according to NASA, the images suggest what we would have seen if only we could have gone along truly, instead of vicariously, for the ride: this is a landscape both ancient and modern, already known and appropriated by humanity in and with the names of its dreamers and explorers. But what did we really find?

The Beagle Hasn't Landed

While America projected landing its two Mars Rovers in January, 2004, the European Space Agency (ESA), as part of its Mars Express mission, planned a landing of their own for December, 2003. The English leaders of the lander component of the mission named their creation "Beagle 2," honoring the ship on which Englishman Charles Darwin had sailed as he formulated his theory of biological evolution. Beagle 2, like the American Rovers, was to explore Mars for signs of life. The images here from the ESA, however, do not project typical American, or at least NASA, attitudes toward exploration. The caption for the left-hand image was, "When folded up, Beagle 2 resembles a large pocket watch." Judging by the enormous grin on the scientist's face, this would be the best Christmas present ever. The right-hand image is an ESA artist's rendering of the unfolded craft as it should look on Mars, reminding us, perhaps, of the parts of a toy spread on the carpet waiting for an adult to supply "some assembly."

NASA, however, provided news agencies with a more lurid rendering of Beagle 2 in action. The image worked much better to stoke the drama of a space race. The *Ann Arbor News* (December 18, 2003) printed it with an ar-

ESA

ticle headlined "Mars failure is never far away" that carried a subhead reminding us that "Several probes are scheduled to drop by the 'death planet.'"

Although the Mars Express mission returned a great deal of useful information from the instrument capsule it put in orbit around Mars, "unfortunately," as an ESA news release stated without apparent emotion, "the Beagle 2 lander was declared lost after it failed to make contact with orbiting spacecraft and Earth-based radio telescopes."

It turned out that the Americans were right: This was a death planet for the beagle. Would a Rover do any better?

A-Roving We Will Go

NASA/JPL

Spirit landed successfully on January 4, 2004. The balloons that cushioned it were deflated and fell away. As soon as possible, NASA issued this 360° mosaic of Spirit on station, clearly functioning since only Spirit was there to take and transmit the images that went into this composite. Opportunity was similarly successful, elsewhere on Mars, on January 25, 2004. On the 26th, the *New York Times* reported the jubilation at NASA.

> This is exactly what it looked like in my wildest dreams," said Dr. [Steven W.] Squyres, a professor of astronomy at Cornell University who is principal investigator for the two NASA rovers currently on Mars. . . . "I will attempt no science analysis, because it looks like nothing I've ever seen before. . . . I've got no words for this.

The multiple self-contradictions in Dr. Squyres' statement are entirely understandable, of course, as a reflection of his well earned exuberance. But we should also note, in the reporter's language that doubtless reflects the atti-

NASA/JPL/Cornell

tude projected officially by NASA, the reference to the "two NASA rovers *currently* on Mars." Even at this initiating moment, the thought of returning for more already shapes the discussion. As Dr. Larry Soderblom is reported in the same article to have said, the landing site is "'Martian pay dirt.'" Was it only coincidental that that term means "find the gold" and "strike it rich"? Dr. Soderblom is a geologist.

The rovers were remarkably successful in many regards, sending back a steady lode of areological data of all sorts, including photos for both analysis and display. On June 30, 2004, the *New York Times* printed this image. At first glance, given the lack of scale, it appears to be some automated mining operation of the sort we might expect in the American West, a hill-clearing enterprise to raise passions both commercial and environmental.

The headline does not disabuse this first impression: "Taking a Bite At 'Breadbox.'" Reading further, we discover that this is the rover Spirit at work, which implies that the pun on "bite" and "breadbox," suggesting the consumption of the substance of the landscape, is not a matter merely of the headline writer's imagination. It was someone at NASA, not the reporter, who decided to call the hillock "Breadbox."

The *Times* article begins, "This ground-eye view of the invasion of Mars shows the rover Spirit reaching out to inspect a rock called Breadbox in a region of the planet know as Columbia Hills." At the end of *The War of the Worlds*, Wells's narrator wonders whether some future conflict between Earth and Mars will see triumph for them or for us. In the newspaper lead, "war of the worlds" and "invasion from Mars" are neatly tied together and turned around as "the invasion of Mars" by us. Not only have the Americans named one feature after a domestic contraption that feeds us, but the entire region is named after a synonym for America. Will Americans in fact become as exploitative as Bradbury's early colonists were in fiction, or will the millennia of imagination serve as prelude to a future of real glory?

INDEX

About the Author

ERIC S. RABKIN teaches in the Department of English at the University of Michigan. He is the author (and editor) of more than thirty books on science fiction and writing, including *Science Fiction: A Historical Anthology and the Fantastic in Literature*, and more than 100 articles in scholarly and mainstream media.